HACK THIS:

24 INCREDIBLE HACKERSPACE PROJECTS FROM THE DIY MOVEMENT

JOHN BAICHTAL

800 East 96th Street
Indianapolis, Indiana 46240 USA

HACK THIS: 24 INCREDIBLE HACKERSPACE PROJECTS FROM THE DIY MOVEMENT

ISBN-13: 978-0-7897-4897-3

ISBN-10: 0-7897-4897-5

The Library of Congress Cataloging-in-Publication Data is on file.

Printed in the United States of America

First Printing: October 2011

TRADEMARKS

WARNING AND DISCLAIMER

BULK SALES

Que Publishing offers excellent discounts on this book when ordered in quantity for bulk purchases or special sales. For more information, please contact

U.S. Corporate and Government Sales

1-800-382-3419

corpsales@pearsontechgroup.com

For sales outside of the U.S., please contact

International Sales

international@pearson.com

EDITOR IN CHIEF
Greg Wiegand

ACQUISITIONS EDITOR
Rick Kughen

DEVELOPMENT EDITOR
Rick Kughen

TECHNICAL EDITOR
Mark Reddin

MANAGING EDITOR
Sandra Schroeder

PROJECT EDITOR
Mandie Frank

COPY EDITOR
Megan Wade

INDEXER
Lisa Stumpf

PROOFREADER
Leslie Joseph

PUBLISHING COORDINATOR
Cindy Teeters

DESIGNER
Anne Jones

COMPOSITOR
Studio Galou, LLC

CONTENTS AT A GLANCE

TABLE OF CONTENTS

ABOUT THE AUTHOR

John Baichtal is the founding member of Twin Cities Maker, a hackerspace organization that has been collaborating for almost two years. Twin Cities Maker has its own rented warehouse, the Hack Factory, complete with a welding station, a woodshop, a classroom, and an electronics area. John is currently writing *The Cult of Lego*, a book about adult Lego builders for No Starch Press. He has written dozens of articles for print, including pieces for *MAKE Magazine*, *Kobold Quarterly* (a D&D magazine), and *2600: The Hacker Quarterly*. He has blogged for Wired.com (*GeekDad* blog) for four years and *Make: Online* for a year, with more than 1,000 posts published during that time.

DEDICATION

To hackers everywhere, for expanding the realm of the possible;

to Eileen Arden, Rosie, and Jack, for their limitless interest in the mysteries of the world; and to Elise, for making it all worth it.

ACKNOWLEDGMENTS

I want to thank Gareth Branwyn for his assistance and encouragement in this project, and to all the hackerspace members who contributed projects and photos to this book.

WE WANT TO HEAR FROM YOU!

As the reader of this book, *you* are our most important critic and commentator. We value your opinion and want to know what we're doing right, what we could do better, what areas you'd like to see us publish in, and any other words of wisdom you're willing to pass our way.

As an editor in chief for Que Publishing, I welcome your comments. You can email or write me directly to let me know what you did or didn't like about this book—as well as what we can do to make our books better.

Please note that I cannot help you with technical problems related to the topic of this book. We do have a User Services group, however, where I will forward specific technical questions related to the book.

When you write, please be sure to include this book's title and author as well as your name, email address, and phone number. I will carefully review your comments and share them with the author and editors who worked on the book.

Email: feedback@quepublishing.com

Mail: Greg Wiegand
Editor in Chief
Que Publishing
800 East 96th Street
Indianapolis, IN 46240 USA

READER SERVICES

Visit our website and register this book at informit.com/register for convenient access to any updates, downloads, or errata that might be available for this book.

Foreword

SUPPORT YOUR LOCAL HACKERSPACE

At Maker Faire Detroit in the summer of 2011, I found a maker exhibit inside the Henry Ford Museum where a medium-sized helium balloon was rising and falling gracefully. When I talked to the group of makers, I learned that the five of them were members of the LVL1 Hackerspace in Louisville, KY. Their project was called White Star Balloon (whitestarballoon.com). As they told me proudly, the goal of their project was to accomplish the first successful trans-Atlantic crossing by a small robotic weather balloon. With perfect irony, they explained that the name "White Star" came from the shipping company that built the Titanic. As makers, they anticipated some failure before eventually succeeding.

As much as the project itself excited me, I was happy to know that this project had come from the LVL1 hackerspace. In 2009, three people from Louisville, two of whom were teachers, came to the Maker Faire Bay Area where they first learned about hackerspaces and interacted with others who had organized them. Immediately, they thought Louisville should have its own hackerspace, and by July 2010, it did. Soon LVL1 became a gathering place for a number of people with various levels of technical expertise, many of whom did not know that others like them existed. One of them was Dan Bowen, a ballooning enthusiast who worked by day in R&D at an autoparts maker. At LVL1, Dan found others who shared his interest and were willing to contribute in unique ways to this ambitious project. Some were students at the university; others like Bowen had day jobs but wanted to be part of something new and fun. None of them were the founders of the LVL1 hackerspace but the space made the project possible.

I enjoyed talking to the White Star Balloon team about various electronic systems they built for the project. Its automated ballast system, which normally was a soda bottle filled with alcohol, had to be modified for Henry Ford Museum to work with BBs instead. The balloon I saw rising and falling was doing so under the control of an Arduino microntroller, which they had programmed. They were a team of experimenters—smart, enthusiastic and determined. They were self-organized, self-funded, and self-motivated. They were exactly the kind of people and projects that I wanted Maker Faire to discover and celebrate.

The White Star Balloon project is just one example of what we're seeing from hackerspaces. In areas like Detroit, we've seen new hackerspaces become a reality in just the last year or two. I3 Detroit has 70 members, many of whom have donated tools to the space, which itself continues to expand. OCD is a hackerspace located in Eastern Market, a former meatpacking area that is now the site of a weekend farmer's market. The Mt. Elliott Makerspace is located in the basement of a church and serves neighborhood children who are learning to solder and make musical instruments. On a Saturday morning at Eastern Market, the kids from the Mt. Elliott Makerspace were teaching teenagers and adults how to solder, right next to summer tomatoes and cantaloupes.

All of these hackerspaces reflect a new DIY mindset that will be reshaping culture, commerce, and community in the coming years. I call it the Maker Movement, which has grown out of *Make* magazine and Maker Faire. Its basic premise is that what you do matters, and sharing what you do matters even more. Exploring your interests connects you to others and forms the basis for community, which is at the center of a hackerspace. It's important that we see new hackerspaces forming in places that don't have them and that we attract all kinds of people to hackerspaces.

However, there are many people who do not understand what a hackerspace is and have not had the chance to talk with hackerspace organizers. If I were to ask the average person on the street about a hackerspace, I'd likely get a puzzled look or a frown. I'd probably have to explain that these aren't hackers who are stealing credit cards or cracking into government computers, although I'd have to admit that some of them might know how to do that. I'd try to explain that there are over a 100 hackerspaces in the United States alone. I might try to describe a hackerspace as a clubhouse, although I worry that they might think of the Elks Lodge or Odd Fellows Hall of yesteryear. After all, the point of a hackerspace isn't just getting together.

It's about learning to do things and making things for all kinds of reasons. I might refer to the 70s era Homebrew Computer Club in the Silicon Valley where enthusiasts gathered to share the designs of computers they were building. Out of these meetings grew an entire personal computer industry. I might mention that we've already seen a 3D printer kit, the MakerBot, originate from the NYC Resistor. I'd like people to understand that hackerspaces are becoming centers for innovation, creativity, and collaboration, and they can be located anywhere from a church basement, a rustic shed, or an old factory.

My hope is that this book by John Baichtal will introduce more people to the practices of hackerspaces. Understanding hackerspaces requires more than an explanation, however. John gives you a sense of what people do in a hackerspace with examples of real projects, which range from the practical to the playful. Each of his chapters profiles the kind of creative work you'll find almost nowhere else but a hackerspace.

Less and less do you find these kind of exciting projects inside of universities or corporations. Perhaps the stakes are too high in those institutions to foster this kind of creative freedom. Sadly, many of our schools are not providing any level of hands-on technical education. Hackerspaces represent a sign of hope for the future. Loosely organized and non-hierarchical, hackerspaces can develop the talents found in a community and break down the barriers to participation in technical fields. And while there may be an adhoc global network of hackerspaces, each hackerspace does its own thing, locally, and that's really important. Each hackerspace is an experiment, and we need lots of experiments.

I hope you will be inspired to get involved in an existing hackerspace or to help start a new one. We can't predict the downstream effects of creating a local hackerspace, but we do know that opening the door to a hackerspace is an invitation for new projects like White Star Balloon to become a reality in your own community.

Dale Dougherty
Founder, *Make* magazine and Maker Faire
O'Reilly Media

HACKERSPACES: THE BLEEDING EDGE OF THE DIY MOVEMENT

Imagine a friend calls you up on a Wednesday night and invites you to the "open night" at the local "hackerspace." You have nothing going on so you agree to tag along and find out what it's all about. You and your friend arrive at a small building in an industrial part of town. The front door is propped open, welcoming visitors and letting the sights and sounds from the building's interior out into the neighborhood.

You step through the doorway and into a noisy room full of people and equipment. Everywhere you look there are men and women doing interesting things. At one table, a couple of guys are hunched over a 3D printer that is outputting in plastic, layer by layer. As you watch, a sprocket begins to take form on the printer's build platform. In the wood shop a woman is milling hardwood for a suite of custom furniture she's designing, peering through her safety goggles at a table leg spinning in a lathe. Other members are playing around with a CNC machine, a computer-controlled milling machine that grinds away at a block of wood or metal, while others are soldering electronic components together. Still more members are collaborating on some project or another, drawing up schematics on a white board.

The collaboration is the aspect of the space that strikes you the most. People are working and talking together. They're sharing information, learning about new things, asking questions, and discussing mutual areas of interest. They're building projects to fill a practical need or simply for the love of it.

This is a hackerspace.

IT'S ALL ABOUT THE COMMUNITY

While they come in many flavors, the typical hackerspace is an organization, like a club, founded by its members for their mutual benefit. They collect dues and rent a storefront or warehouse. They elect officers and buy insurance.

On a very basic level, hackerspaces are simply workshops that are available for rent. Most people don't have the ability to weld up a custom bicycle frame at home, but at a hackerspace, they can have access to a workshop and welding equipment. A clothing designer could use a space's industrial sewing machine, an electronics engineer could troubleshoot a complex circuit with the high-end oscilloscope, and an inventor could prototype a new creation on the 3D printer. All these items could cost hundreds if not thousands of dollars each, but with a hackerspace's members collectively paying for it—$1000 isn't so much when 40 people contribute—these workshops quickly fill up with equipment.

Credit: Anne Petersen

▲ *This lively gathering at Chicago's Pumping Station: One shows the hackerspace scene at its best—creative people sharing ideas in a common space.*

Many hackerspace members are independent businesspeople (crafters, electronic kit sellers, furniture makers, artists, and so on) who consider the subscription model an excellent alternative to outfitting their own workshops, and they can write off their dues as a business expense.

Kellbot, an independent crafter and jewelry maker from New York City, runs a business out of NYC Resistor, a hackerspace in Brooklyn. Primarily, she uses the space's laser cutter, a huge and prohibitively expensive machine that uses a laser to cut through metal, plastic, and wood, following a pattern created on a computer. "I live in the New York City area. Have you seen the size of our homes?" she asked. "Space is the primary reason. Even if I decided it would make economic sense to have my own laser cutter, there's nowhere to put it."

Some hackerspaces are simply subscription-based workshops available to any who can pay the fee. The most notable of these is the for-profit chain of TechShops, with franchises located in San Francisco and Raleigh, NC, as well as others planned in four more cities. If you simply want to swing by and laser-cut a wooden puzzle or solder a LED lamp for your house, you can buy a day pass at a TechShop or similar service for a small fee or buy a monthly membership.

However, many people feel these spaces miss out on something: a community. Kellbot listed the space's community as a leading reason for working there. "The people are what make it. One of the things that sucks about running a business by yourself is that it can be very lonely."

Mike Hord, founding president of the Hack Factory in Minneapolis, agreed. "A hackerspace collective is a different mental model than a TechShop, in much the same way as a farmer's market is from a grocery store," he said. "Sure, the superficial trappings are similar—tools for use or food to purchase—but if you've shopped at a farmer's market, the difference is immediately apparent: people sitting, having coffee, chatting. In a word, *community*, which is the key to a hackerspace being successful."

Surrounding one's self with a large group of talented and engaged people can inspire a member to tackle great challenges. "We all need community," said Mitch Altman, cofounder of the San Francisco hackerspace Noisebridge. "One of the reasons we survived as a species is that we evolved over millions of years to get together in community to support one another. It is in our DNA. And yet, there is precious little of it in our modern world. But we all *need* it."

HACKERSPACE U

Hanging around a hackerspace, you can be exposed to new areas of interest that you never knew you'd like. For instance, an electronics hobbyist may not necessarily know about welding, brewing, knitting, ham radio, but by working alongside other people who are, who knows? That member might discover a new interest. Even better is the opportunity to collaborate on a large project. Each member has an opportunity to learn something new. One collaborator might be an expert programmer, while another might have a degree in electrical engineering. A third member may have nothing to offer besides enthusiasm, an intense curiosity, and a willingness to learn. Together they can create something that individually they wouldn't have had the time, money, or know-how to make.

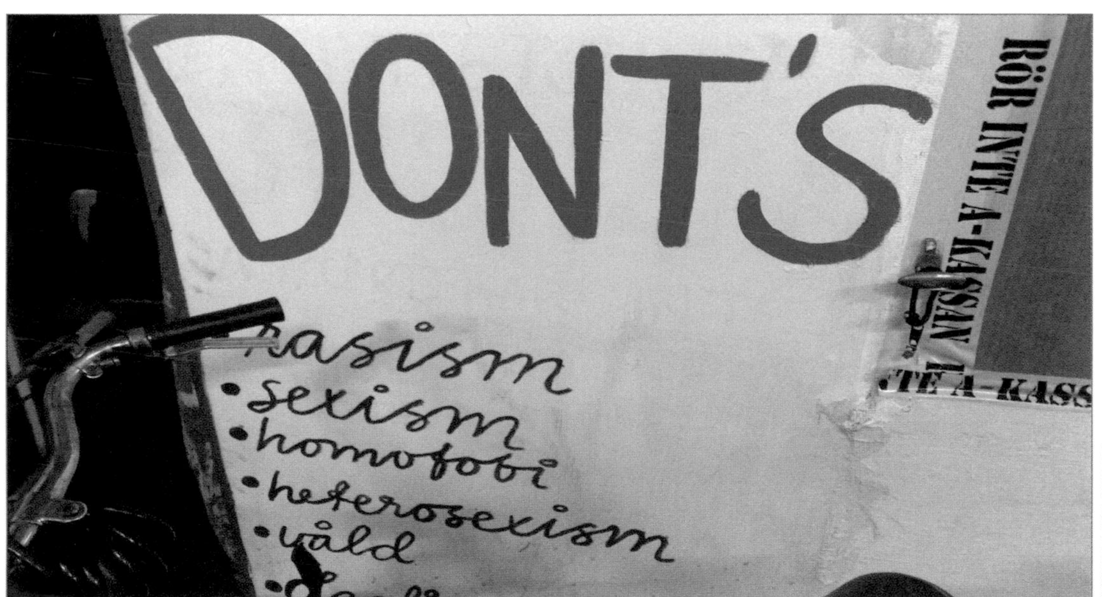

Credit: Peter Rukavina

▲ *This message, found in Malmö, Sweden, hackerspace Forskningsavdelningen, lays the rules down.*

▲ *Hackerspaces are all about tools, and Phoenix Asylum's impressive CNC router is the sort of device sure to attract new members who want to play with it.*

"The projects that hackers work on do have practical applications," HacDC founder Nick Farr said. "But it's not 'work' as much as it is 'play'. We often learn more from things we're engaged with in play. Crossword puzzles engage people in language and symbolic construction; hackerspace projects do the same thing with technical skills, enhancing and learning through play and pure enjoyment."

Most hackerspace members are active learners who don't mind sharing what they've discovered. Hackerspaces are open to all skill levels and any area of interest. If you don't know something, you can go to the space and if no one there knows the answer, chances are people will help you find out. They're a hotbed of doing and learning on par with an academic institution.

Like a school, most hackerspaces offer classes, taught by members or guests. If you're interested in electronics, or knitting, or building furniture, but don't know how, you can take a class at the space and get the 411. These courses typically cost money, with the proceeds benefiting the organization. They cover every topic that could interest a member or potential recruit, usually involving complicated technical matters with a steep learning curve, like programming microcontrollers, the chips found in many electronic projects. Other common topics include programming; soldering; woodworking; as well as training in the use of the hackerspace's high-end tools such as their laser cutter, MIG welder, or CNC machine. Typically, a member must get checked out by a more knowledgeable peer before he's allowed to work with these expensive gadgets.

However, a lot of learning happens even outside of classes. Simply by being in the same room, it's easy for the members to interact with each other and learn about other projects that are being worked on in the space.

Credit: Rose White

▲ *A stenciled message at San Francisco's Noisebridge encourages visitors and members.*

THE FIRST WAVE OF HACKERSPACES

If you've never heard of hackerspaces before, there's probably a good reason: chances are your nearest space didn't exist 3 years ago. In 2005, there were fewer than 20 of these organizations in existence throughout the world, but by 2010 there are over 500 planned or already formed. What happened?

The first hacker collective was the Chaos Communications Club (CCC) in Berlin, Germany, founded in 1981. It began as a computer user's club with only five members and no space, but it continued a slow and steady growth, reaching an incredible 4,000 members by 2010. The CCC is open, anarchic, and wide-ranging in its focus The members describe themselves as "a galactic community of life forms, independent of age, sex, race, or societal orientation, which strives across borders for freedom of information...." There is also an undercurrent of leftist politics that sets them apart from American spaces, which are usually apolitical.

In 1984, the CCC became folk heroes when they performed the first electronic bank robbery and returned the money the next day. A few years later their open and accepting nature burned them when a member used his skills to infiltrate U.S. government servers in an attempt to sell stolen data to the KGB. He was caught with the help of a clever programmer, but the damage was done to the CCC's reputation.

Meanwhile in the United States, a couple of small hackerspaces were founded on a very different model from the CCC. Such programmer hangouts as the L0pht, Hasty Pastry, and New Hack City were private affairs where small numbers of programmers could interact in person. In addition to seldom admitting new members, these American spaces stayed in the shadows as much as possible—the organizations were solely for the benefit of a small and select group of friends.

While this kept the spaces exclusive, it was in some degree a stifling environment when it came to exposing the hackers to new ideas and new skills. "Opening a space for anyone in the community to participate has a natural effect of diversifying not only the interest and the focus of projects in the space," Nick Farr, the HacDC founder, explained. "The best 'hacks' in a hackerspace incorporate many disciplines in ways that are unconventional, help draw in people of diverse backgrounds to share what they know, and find ways to connect." By excluding most people, except for those select friends and colleagues, those first U.S. spaces missed out on the incredible explosion in interest that would follow.

Credit: Marty McGuire

▲ *A class in Pittsburgh's HackPGH highlights hackerspaces' devotion to sharing knowledge.*

HACKERS ON A PLANE

While the first wave of U.S. hackerspaces lurked in the shadows, in Europe, the hackerspace scene was slowly spreading. However, it too could benefit from some new ideas. In 2006, looking to recruit some fresh blood, the CCC invited Nick Farr to Berlin to speak at the Chaos Communications Congress, their biannual club-wide meeting. Farr, who went on to co-found two spaces, was taken to his first hackerspace, a CCC offshoot called c-base, which is laid out on a spaceship theme. "That's really where the whole hackerspace thing hit me," he said. "A real community built around a space that had a strong culture, a huge dance floor/meeting area, a bar, and every kind of workshop space imaginable in the basement. A woodworking shop on one end, a recording studio on the other...and the place was full of people hacking in every way imaginable! I saw it and said, 'This is what we're missing in the United States!'"

Credit: Kim Gillespie

▲ *A member of Louisville's LVL1 works at modifying a toy guitar.*

After seeing c-base, Farr vowed to introduce as many North American hackers as possible to Europe to check out the scene. "I knew that all I had to do was bring people to Germany, show them how the community-oriented hacking worked, and they'd take it home with them," he said. The following year, Farr organized Hackers on a Plane (HoaP), a group plane trip from the Las Vegas hacker convention DEFCON to CCC's camp-out slash hacker con, with a tour of several German and Austrian hackerspaces thrown in.

"What we saw was very compelling," said Mitch Altman, one of the participants. "There were so many amazing people who were presenting so many cool projects, beautiful art, wonderful crafts, creating community, serving community, helping each other, teaching, learning, sharing." The HoaP participants returned home brimming with excitement and the desire to start their own hackerspaces. Within weeks, NYC Resistor, the Hacktory, HacDC, and Noisebridge had been founded by HoaP travelers.

But it wasn't merely Americans who found inspiration from HoaP; the hackerspace scene exploded throughout the world as a result of this new wave of spaces. Within three years, more than 500 new organizations were founded, with hundreds more in the works. They're popping up in cities around the globe, recruiting members and renting spaces. What will come of this phenomenon? No one knows, but it's an incredible and utterly amazing experience.

THE FIRST HACKERS

The etymology of the hackerspace confuses some people, who associate the term *hacker* with computer criminals. However, the truth is far more complex and a hell of a lot more interesting.

Hackers got their start long before the dawn of the Internet. In the 1950s, students at the Massachusetts Institute of Technology (MIT) formed the Tech Model Railroad Club (TMRC) with the goal of building progressively bigger and more complicated track layouts. Whole legions of students had participated in the TMRC during their time at MIT, and the club's layout had grown to incredible sophistication, using customized tracks, signaling systems, and landscaping that simply weren't available from hobby stores. The most skilled members of the TMRC called themselves *hackers*, and their ingenious solutions were *hacks*.

What is the precise definition of hack? It's hard to say, but the seminal book *Hackers* by Stephen Levy offers a convincing one: "... project undertaken or product built not solely to fulfill some constructive goal, but with some wild pleasure taken in mere involvement, was called a 'hack'.... To qualify as a hack, the feat must be imbued with innovation, style, and technical virtuosity." Notice how there's nothing about committing crimes in that definition? True hacking, the experts agree, is about thinking, and solving, and learning, not about taking or hurting.

"A *hacker* is, quite simply, someone who messes around with stuff," Emmanuel Goldstein said. Goldstein, the editor and publisher of seminal hacker publication *2600*, has long been in the forefront in the battle to educate the public on the true meaning of the term. "It doesn't have to be a computer. It doesn't even have to be technology. A hacker is someone who figures out how to get around barriers, won't take no for an answer, asks an insane amount of questions, and believes in sharing the information he or she discovers."

Indeed, a lot of people who call themselves *hackers* focus on learning to overcome artificial barriers. Urban explorers, who investigate abandoned buildings and map out steam tunnels, are hackers. So are locksport aficionados, who learn to pick locks simply for the love of it. Other flavors of hackers include ham radio enthusiasts, electronic music circuit benders, and phreaks (who explore the world's phone networks as energetically as urban spelunkers find a city's forgotten spaces).

This spirit of adventure, eagerness to solve problems, and joy of collaboration shone clearly in those original model railroaders at MIT, and these qualities did not go unnoticed by the faculty. One day the school got a computer, a TX-0 on permanent loan that wasn't powerful enough to interest the professors and graduate students, who had access to the school's top-of-the-line IBM. But it was certainly far cooler than anything the hackers of TMRC had ever gotten their hands on. Soon the railroaders began sopping up every spare minute of the machine and began learning how to program it. In time, as their skills became recognized by the faculty, they were recruited to serve as administrators of the TX-0. They explored its capabilities to the fullest, writing the first digital music and creating simple games. They competed with each other to code the leanest programs—those that used the fewest lines of instructions. From that point forward, hackers were associated with computers, but that field has never been the sole focus of the hacker, as the hackerspace scene proves.

HACKERS AND CRIME

So how did hackers get the reputation for committing crimes? As computers became household items in the 1980s, the culture of hackers at MIT and other institutions filtered down to these first consumers. There was still a spirit of exploration. After all, these operators of home computers were pioneering a new technology the same way the model railroaders learned to hack the TX-0. But there were growing pains.

Part of it might have been that the culture of hacking and the concept of a corporate marketplace didn't precisely synch up. Proprietary ownership of software always baffled hackers. A program is simply a series of instructions, so how can someone say that no one else is allowed to instruct the computer the same way? And so hackers learned to crack software protection and "improve" upon the program, usually by making it freely available for others to use. Similarly, those first system operators deliberately eschewed security protocols, leading to a culture where everything from server logs to databases and email was open and available for anyone to inspect.

It could have been the natural adventurousness of kids, learning how to break into university or corporate computer systems simply to find out what was there. Teenage coders broke the copy protection on video games they never could have afforded. Phreaks hacked conference call services to hold impromptu meetings late at night and soldered "blue boxes," which simulated the secret tones that phone companies used to regulate long distance service to make free calls.

In time, the mainstream community began to associate computer hobbyists with crime, encouraged by news reports and movies on the lookout for more villains. They fed into a sense of unease about the growth of the

Internet, focused on the hacker as the principal antagonist of the information superhighway when they were arguably its heroes.

PRANKS

Perhaps another reason hackers were distrusted was their multigeneration love affair with pranks. Long before there was a TMRC, there was a rich tradition of pranks at MIT, and in the school's unique jargon these were called *hacks*. On a very basic level these hacks were on par with traditional college antics like TPing the sorority house, but they were far more clever. For instance, to celebrate the 40th anniversary of the Apollo moon landing, MIT hackers put a replica of the Eagle lander on top of the school's Great Dome. In 2000, hackers dropped stuffed beavers from a hot air balloon onto a Harvard-Yale football game.

Even after hacker culture filtered into the mainstream, the pranks continued. Rob T Firefly, a hacker affiliated with a group of phreaks called the Phone Losers of America (PLA), described the unique interweaving between the concept of the hack and the prank. "I'm a prankster and a hacker, pursuits which are far more related to each other than one might think," he explained. "As I see it, the best pranks have always been about altering the context of normal situations in funny and creative ways; the target expects a situation to go one way, but it

Credit: Mitch Altman

▲ *A soldering class in San Francisco hackerspace Noisebridge introduces newbies to one of the most important hacker skills.*

unexpectedly goes somewhere else entirely in a funny way."

The art of a hack, or prank, is to accomplish something in an unexpected and interesting way, without much concern over any preexisting rules or expectations dictating how they should go about it. PLA has performed such mischief as splicing 900-line psychics and sex workers into the same call and changing the signs of a McDonald's restaurant to show subversive messages. Even other hackers are the target of their pranks. After the PLA released a popular video showing them harassing fast food drive-through cashiers by tapping into their radio communications, they included fake DIY instructions that had people taking apart their toasters and other household appliances in search of mythical crystals they allegedly needed to recreate the trick.

Rob described the intersection between a hack and a prank: "A good hack is basically a successful prank played on common expectations, just as a good prank is basically a hack on society."

A LOVE OF COMMUNITY

There are two factors, beyond a love of technology, that shone through the computer culture from the TMRC days to today: a love of collaboration and a desire to freely share their knowledge. The TMRC members shared their discoveries with one another, but this was a cinch because they were mostly friends and classmates and worked in close proximity with each other.

The philosophy was better tested in the '80s and '90s when the Internet was gearing up and the hacking community was spread thin. Unable to collaborate in person, hackers exchanged textfiles of lore, usually truthful but sometimes hilariously inaccurate. New publications like *Phrack* and *2600* served as more formalized textfiles. Online hacking groups like the PLA were organized, and some of these took on the notoriety of criminal gangs, though actual victims were few and far between.

Even these attempts at community weren't enough, however. Something was missing—the collaborative environment that comes only with being in the physical presence of one's peers. There was an attempt to get people out of their houses and away from their computers. In 1987, *2600* publisher Emmanuel Goldstein organized the first 2600 meetings, hacker gatherings that he stipulated must be welcome to all comers and had to be held in public. Primarily this was to help combat the image of the hacker as a lone criminal, but it also let people with similar interests interact with one another in person.

Similarly, in the late '80s and early '90s, the first hacker conventions were organized, and early events included SummerCon in 1987, HoHoCon in 1990, DEFCON in 1993, and Hackers on Planet Earth in 1994. A hacker convention is an exciting time when friends who know each other only by a screen name gather to listen to lectures, participate in contests, pull pranks, and socialize. It's also a time to learn new ways to explore the world, with discussions of food chemistry, urban exploration, and microcontroller hacking.

A lot of people said, 'Why can't we have DEFCON all the time?'" Nick Farr said. "The hackerspace is a direct answer to that call. The ability to share tools, resources, work out real-world problems, and have a space for lots of groups to meet has brought the community together in ways I couldn't have imagined and really answer the question as to why we need hackerspaces."

Credit: Matt Joyce

▲ *NYC Resistor member Zach Smith fabricates a replacement part for an ancient teletype machine.*

Today, those dreams are realized, as hackers have come together to explore and learn and be around people like themselves. The hackerspace phenomenon is on the upswing, with hundreds of spaces being founded every year. The energy of these spaces is exhilarating. The creative powers of so many smart people, the eagerness to share knowledge, and the ability to stretch the limitations of our everyday lives has caught the imagination of thousands of people around the world.

Welcome to the hackerspace.

PROJECT 1

KARATE CHAMP GAME

The Hack Factory built a video game without the video—a mechanical game hailing the kung-fu arcade games of yore. Instead of moving pixels, however, pressing buttons commands humanoid robots to do battle.

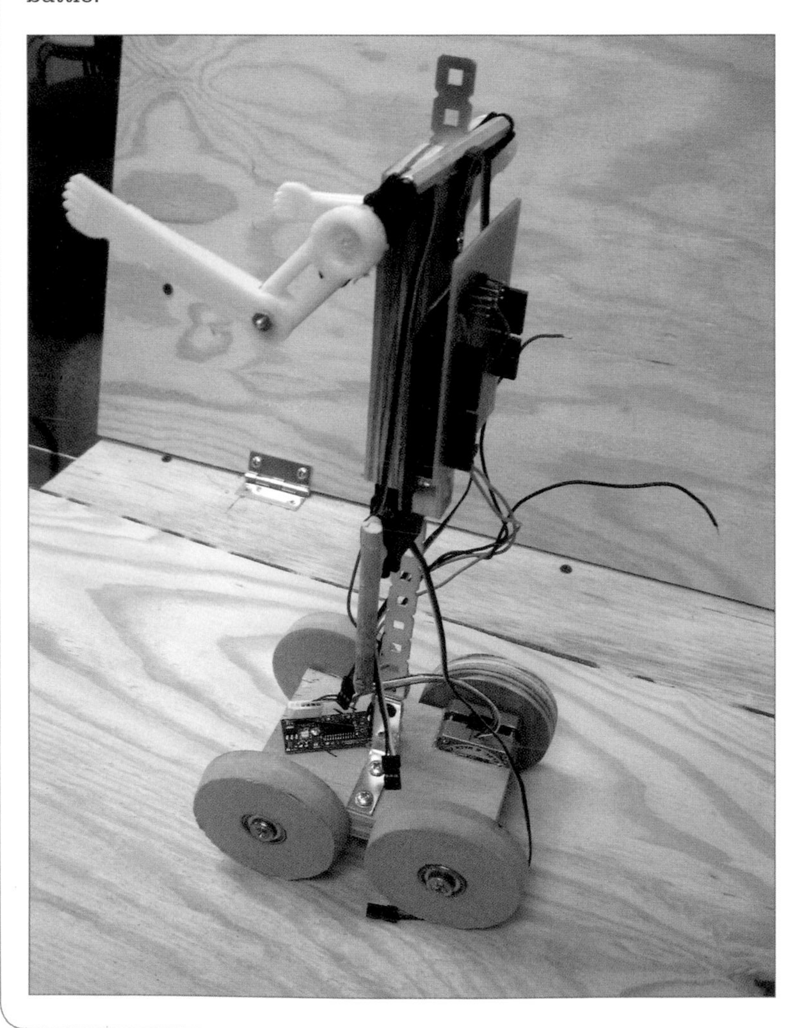

HACKERSPACE PROFILE: THE HACK FACTORY

Location:

The Hack Factory

3119 E. 26th Street

Minneapolis, MN 55406

www.tcmaker.org

Organization type: Minnesota nonprofit, 501(c)3 status pending

Founding date: January 2010

Number of members: 40

Dues: $50/month full membership, $25/month student or starving hacker

Size of the space: 7,800 sq. ft.

Officers/Leaders:

- Dave "videoman" Bryan, president
- Wayne "wammie" Martinson, vice-president
- Bob "ObiBob" Poate, secretary
- Brandon "orion" Paplow, treasurer
- Riley Harrison, board member
- Karin Fitchett, board member
- Michael "metis" Freiert, development coordinator

Notable equipment: Complete wood and metal shops, electronics test equipment and soldering stations, craft area, MakerBot, CNC routers

THE HACK FACTORY SPACE DETAILS

Twin Cities Maker/The Hack Factory is a group created by two organizations that joined together to rent space for a workshop. They organization is called Twin Cities Maker and the facility is the Hack Factory.

"The space is basically a long warehouse space that we are subleasing," president Dave Bryan explained. "Many of the members have put a lot of time and effort into making and creating the space and having a place for everyone to collaborate."

The space is shared with a commercial security company, so it's super-secure with motion sensors and multiple security cameras. The downside of that arrangement is that the Hack Factory has to keep certain areas of the space clear for the landlord; also, nonmembers walk through the space every day.

Credit: Paul Sobczak

▲ *The Hack Factory's low-key exterior; the organization sublets from another company.*

THE PROJECT

Participants: John Baichtal, Adam Wolf, Jon Barclay, Karin Fitchett, and Jude Dornitch. *Note: This project was co-created by the author, who is a founding member of the Hack Factory.*

Adam Wolf and I wanted to build something fun for the Hack Factory, some sort of interactive furnishing that would give visitors and members something to do when they weren't hacking. We decided to build an arcade-style fighting game like the original Karate Champ—only with robots instead of shapes on a screen. Scoring works simply, with LEDs indicating remaining "life" and flashing bright red indicating a hit. A simple, Arduino-based music player provides sound effects over desktop computer speakers. (See the Tool Bin sidebar to learn more about Arduinos.)

Credit: Paul Sobczak

▲ *The Hack Factory's huge wood and metal shop dominates the facility.*

Credit: Paul Sobczak

▲ *Members and visitors interact in the Hack Factory's social area, called the Family Room.*

Credit: John Baichtal

▲ *Karate Champ comes together with the LED strip powered up and one of the robots ready for action.*

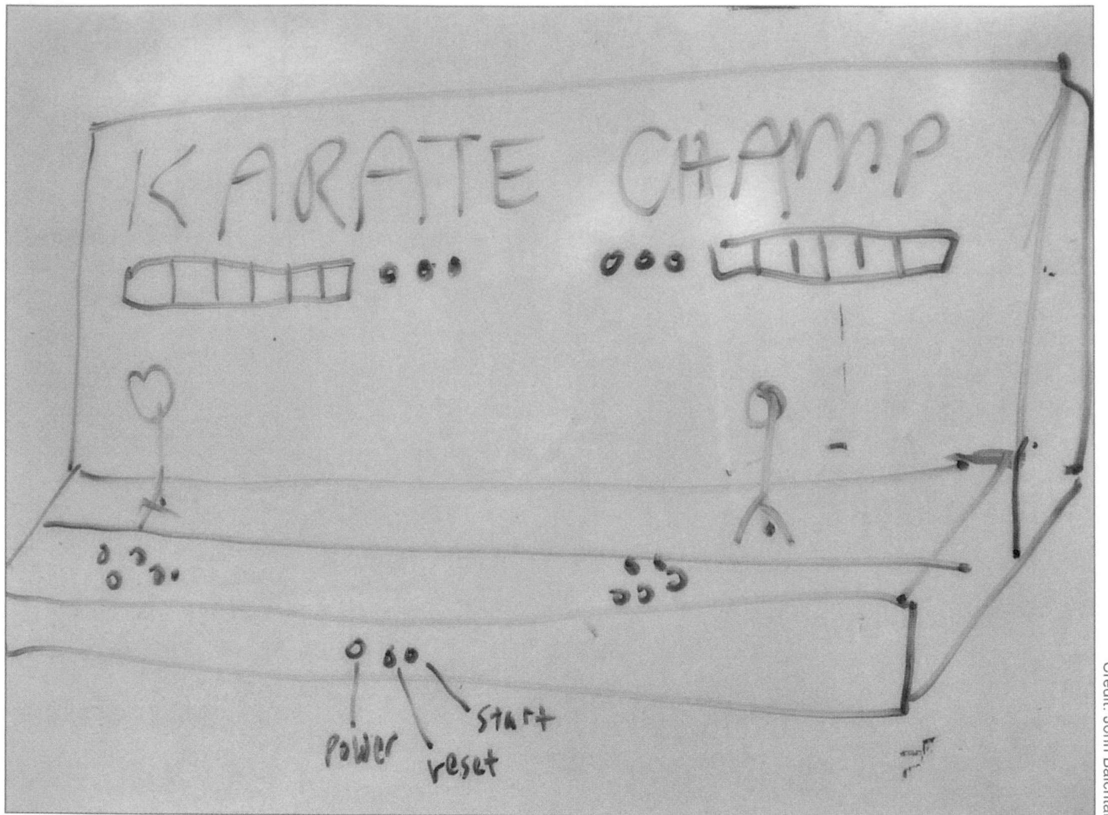

Credit: John Baichtal

▲ *A white board concept of the game.*

The first step was to design the robots, the thought being that they were the most important and complicated part of the machine. We finally decided on robots that could punch with either hand and kick with one leg. We also envisioned a "block" maneuver where both arms rose up to protect the head and torso. The robots wouldn't actually punch each other; the hits and kicks would be purely cosmetic and the game resolution would take place within the software.

Looking for parts, I grabbed a flatbed scanner from the Hack Factory's junk shelves (see sidebar) and disassembled it, salvaging two stepper motors that you'd probably find in a store for $20 each. Adam and I decided that these would be the motors that would move the robots back and forth.

The punching and kicking motions were accomplished with servos; we tried out a couple kinds of hobbyist motors but ultimately bought six HiTechnic HS-81s with a very quick response time but relatively low torque. We wanted speed and didn't care about torque because the servos wouldn't be lifting anything but their own limbs. (For an explanation of stepper and servo motors, see Project #2, "Sudo Make Me a Sandwich Robot.")

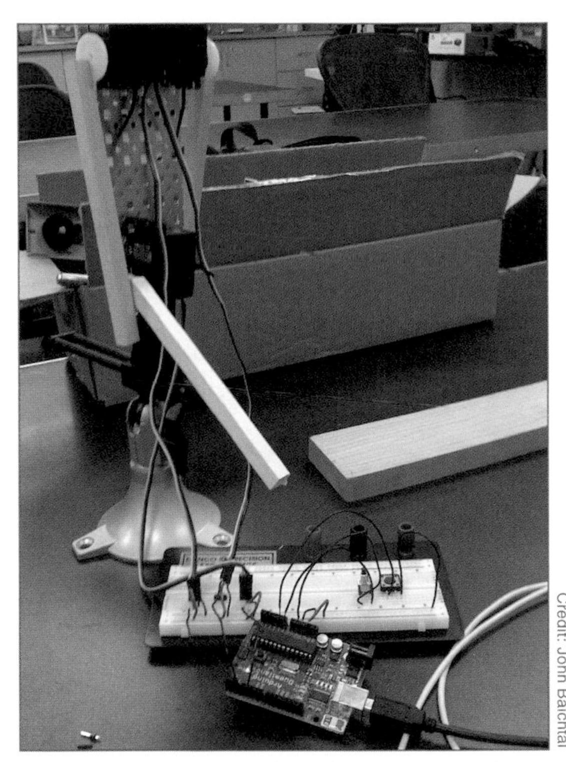
Credit: John Baichtal

▲ *An early prototype of the fighting robots shows their arms and single moveable leg.*

Credit: John Baichtal

▲ *After attempting to 3D print wheels, we eventually decided to cut them out of wood.*

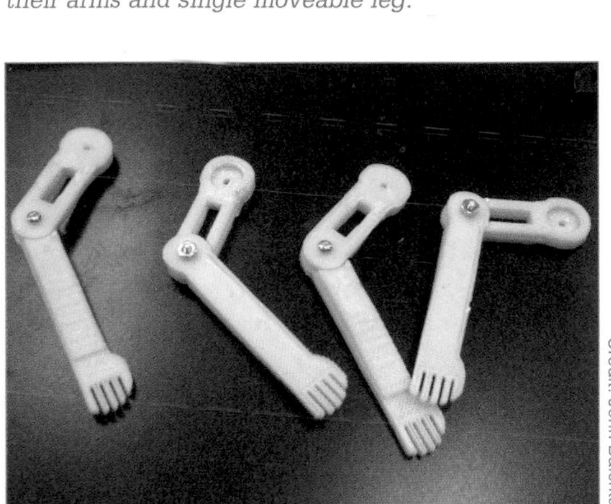
Credit: John Baichtal

▲ *The robots' arms were designed digitally and printed on the Hack Factory's 3D printer.*

The robots were designed to move around on small wheeled platforms that would be concealed beneath the game surface. Each robot packed an Arduino, a prototyping board designed to manage each robot's tangle of wires, as well as a motor-control circuit board. Each bot would trail behind a slender power/data cable. We used scrap wood for the robots' bases, with hardware store angle braces and girders from a VEX building set forming the chassis. We used a hole cutter to cut the wheels out of 5/8" MDF, with skate bearings helping them spin. Meanwhile, our friend Jon Barclay designed and printed arms on the Hack Factory's 3D printer. (See Project #5 for more information about this technology.) Each robot also has a pair of bump

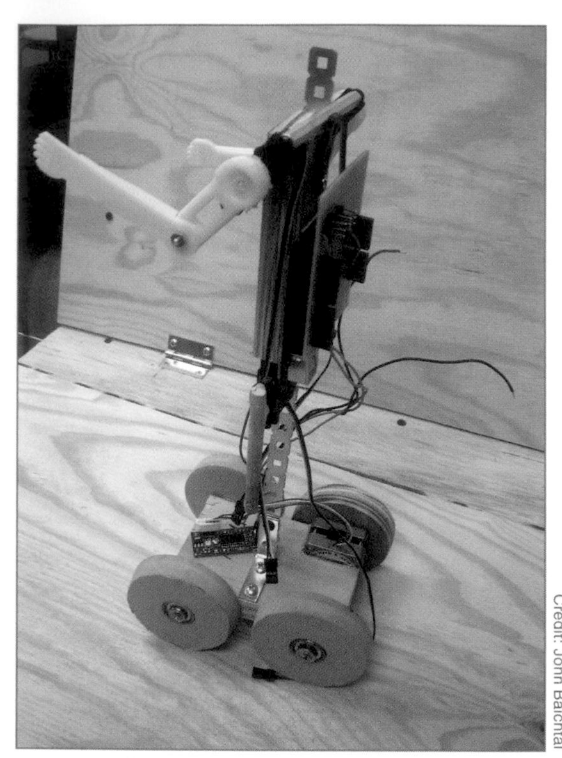

▲ *The fighting robots are designed to move on small wheeled carts that are invisible during actual game play.*

▲ *The brain of the game is an Arduino with a music-playing add-on board and a piece of perf board used to manage the various wires.*

switches on its base so the game can reset them as needed—simply back up the robots until their rear switches are triggered. The front switches tell the game that the robots are in contact with each other.

Next we worked on the game cabinet. We wanted it to be playable on a table, so we needed something relatively simple—basically, a wooden box with a track for the robots and a marquee to display the score. I bought a 4"×8" sheet of 1/4" plywood from the hardware store and cut it up on the hackerspace's table saw, forming the pieces into the game console using angle braces and screws. I designed the top of the console to flip up on a hinge so we could access the wiring inside. I also made the marquee fold down so the game could be stored more easily. The whole thing sits on little rubber feet so the table doesn't get scratched up.

Credit: John Baichtal

▲ *Project participant Adam solders components for one of the robots.*

Credit: John Baichtal

▲ *The underside of the arcade controls shows them wired to the MCP.*

Credit: John Baichtal

▲ *The game console with its arcade controls installed.*

Meanwhile, Adam, who had constructed a video game cabinet in college, bought arcade-style buttons and joysticks to give the players a more authentic experience. He also picked up an LED strip for the marquee that consists of 32 individually addressable RGB lights, meaning that with the help of an Arduino, we could control every one of those LEDs either by itself or with others and could make it any color we wanted. The LED strip would display the score, provide fun animations, and flash red to indicate a hit.

I mentioned that each robot would have an Arduino onboard, but we also needed a central Arduino to manage sounds and score. We dubbed this one Master Control (a reference to the 1982 movie *TRON*); it's an Arduino Mega with a sound-managing add-on board called a Voice Shield as well as a scrap of perf board used to manage all the wires leading from the arcade controls. The whole shebang is powered by a commonplace PC power supply.

The Arduino code makes the whole project work. The entire game is run from the central Arduino, with the robots' onboard Arduinos managing their movement as well as the punching, kicking, and blocking motions. With game resolution decided by software

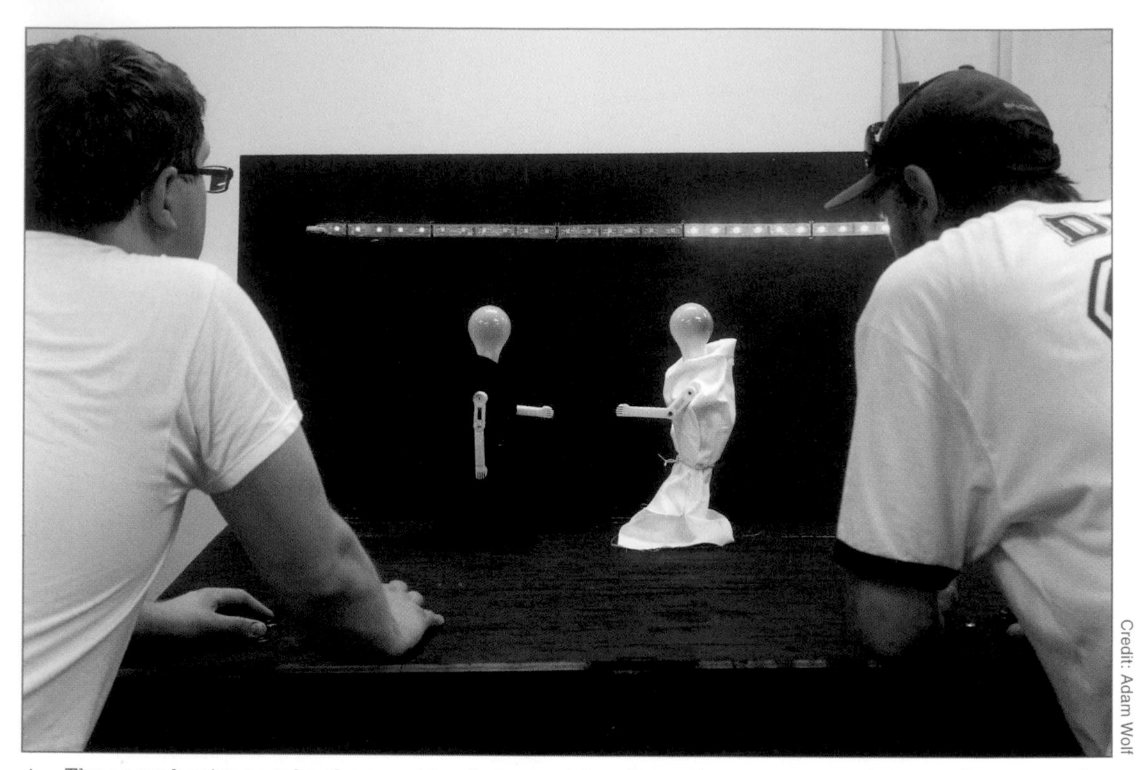

Credit: Adam Wolf

▲ *The game begins to take shape, with robots receiving decorative costumes and the scoring system in place.*

rather than some physical action, we wanted it to be as realistic as possible. A successful hit rocks the other robot back a small distance, and if it is backed up all the way to its end, the fight is basically over. We thought about programming in quirks, like making one robot slightly better at kicking and the other slightly better at punching, but ultimately we made them identical.

When you power on the game, it immediately resets itself by reversing the steppers on both robots until they are backed up all the way and their bump switches trigger. Then the fight begins, with each bot getting 10 "hit points." The robots battle back and forth until one wins, and the software makes his arms wave excitedly in the air while the LED strip flashes. Victory!

Adam and I enjoyed building the game and learned a lot in the process. Even better, the hackerspace gained a new toy that we can show off to visitors and even bring to local events.

JUNK

One person's junk is another person's treasure, the saying goes, and that's definitely true in the hackerspace. Say you have an electronic gadget: a toy, a DVD player, a boom box. It may still be operational or it may not; either way, you don't want it anymore. You could throw it away, or you can find a home for it on a hackerspace's junk shelf.

The junk shelf is where hackers go first when they need a part. You'd be surprised what you can find: the average flatbed scanner has easily removable parts like stepper motors and gears. Whole boxes of electronic components might be as pristine as the day they were manufactured—or they might be totally unusable. It's always a crapshoot.

Here's a small sample of the equipment found on one shelf at the Hack Factory:

- Old printers
- Two Ni-MH batteries from Ryobi drills
- A VCR
- Compartmentalized boxes full of mysterious components
- Power adapters
- Several speakers
- Numerous spools of wire
- Wrecked RC cars
- Screws
- A stack of obsolete monitors
- A ceramic light fixture
- Three pairs of skis
- Two CPU fans from long-dead Gateway PCs
- Vacuum tubes in original packaging
- A hydraulic jack
- Zip ties
- Fuses
- Spray paint
- Keyboards
- Trackballs
- Darkroom equipment
- Non-code electrical hardware
- Bolts
- Plastic sheeting
- A sink
- A broken chair
- Welders' gloves
- A stack of obsolete tech books

Credit: John Baichtal

▲ *The Hack Factory's junk shelves carry everything from coils of wire to obsolete electronic components to broken toys.*

TOOL BIN

ARDUINO

One of the most important innovations in hobby electronics was the development of an easy-to-master electronic prototyping board called the Arduino.

You could call the Arduino a tiny computer. Like a computer, the Arduino accepts inputs and outputs data. But where a PC might take its input from a keyboard and send data to a monitor, the Arduino is a little more open-ended. "Inputs come in the form of voltages, and the outputs are also voltages," explains electrical engineer and Hack Factory co-founder Paul Sobczak. Input voltages typically come from sensors or buttons, while output voltages usually trigger lights, motors, and other peripherals. And if an Arduino by itself doesn't do what you want, there are dozens of add-on boards called *shields*, adding all sorts of capabilities ranging from Bluetooth communication to playing music to controlling large numbers of LEDs.

Perhaps the platform's greatest asset is its ease of use. It was developed to make electronics tinkering available for artists—Arduino programs are even called *sketches*. The Arduino works with Macintosh, Linux, and Windows computers; it's programmable via USB cable; and it has a massive community of developers who share code and solutions to problems. They're cheap to buy (around $35) and kid-friendly.

While there have been other prototyping boards, none have been as convenient for newbies to learn. "Plug in the board, download the IDE," explains Sobczak, referring to the development environment. "Read great documentation on the Internet if you have questions, and then find out that there is already some code written for you that someone released as open source code."

Laypeople don't need to learn to code; they can simply adapt existing Arduino sketches to their own ends. "It is a common misconception that in order to be considered a coder, you need to be able to write from scratch," Sobczak says. "One of the really great things that the Arduino platform has accomplished is the ability to use existing code that is easy to find and works as a starting base for your project."

Credit: Creative Tools (CC)

▲ *The Arduino has taken countless amateur—and professional—projects from concept to prototype in minutes instead of hours.*

While amateurs love the Arduino, even electrical engineers use them, simply because the board reduces the time between concept to prototype. Sure, they might re-create the circuit later with a custom-programmed microcontroller, but the Arduino lets them test their idea before going to the trouble.

If you're curious about what sorts of projects can be powered by an Arduino, just look through the pages of this book—many of the projects used an Arduino somewhere during the development process or the final execution.

To learn more about the Arduino project and to download the IDE, visit http://www.arduino.cc/.

FAQ

THINGIVERSE

One tool you'll frequently encounter in *Hack This* is the 3D printer, which extrudes layers of plastic and builds up 3D objects, following designs sent from a nearby computer. But where might one get these designs? The best site is called Thingiverse (thingiverse.com), and it encourages visitors to freely upload and download files for 3D-printable objects or, as the site calls them, "things."

All designs can be copied, re-created, or improved upon, making Thingiverse a hotbed for creativity. Even better, if you need a certain part—let's say a gear—there's a good chance you can find it amongst the 8,000 things on the site and print it yourself.

▲ *A printable clock gear from the archives of Thingiverse.com.*

ALT.PROJECT

10X

"Building a functional circuit fosters a sense of accomplishment which is absent from many school curricula," Mike Hord wrote in the project description for 10X, a set of oversized electronic components designed to let kids get hands-on with the technology. The project consists of a prototyping board and components, all about 10 times their actual size—but fully functional. Each simulated component has a real one buried inside of it. Hord and his collaborators at the Hack Factory tested the system with children attending a hackerspace event and were gratified at how excited the kids were to build circuits and learn about what role each component plays.

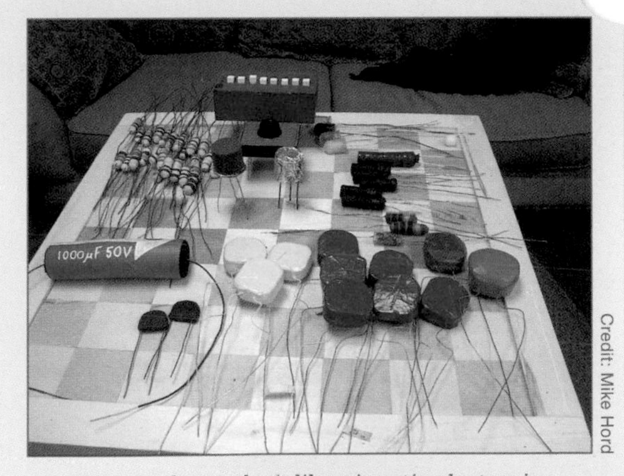

Credit: Mike Hord

▲ *If these shapes look like gigantic electronic components, it's for good reason.*

http://www.element14.com/community/groups/hack-factory-global-hackerspace-challenge/blog/

BUILD IT

Ultimately, the Karate Champ game consists of motorized fighters with software controlling everything and awarding points. The advantage to this model is that the basic format could be changed, even radically, with the software adapted to suit the new architecture. Just follow these directions:

1. Build the fighter robots. We created the robots to be physically identical, but with customization possible on the software end. Each robot consists of four large wooden wheels, an Arduino, 3D-printed limbs, a stepper motor, a controller board, and three servos.
2. Configure the Arduinos. We used one Arduino for each robot and a larger Arduino and Wave Shield to manage the sound and scoring.
3. Build the enclosure. We used plywood for ours, bolted together with L-brackets with hinged access panels.
4. Attach controls. We wanted to offer players an arcade-like experience, so we used standard arcade buttons and joysticks.

5. Add the scoring marquee. The LED strip is a great solution for scoring because each of its lights is individually addressable, meaning that software can control which LEDs light up as well as their brightness and color.

6. Add power. We used a standard PC power supply that easily fit into the enclosure. As a bonus, the power supply's built-in on/off switch serves as a power switch for the game itself. (Not shown.)

7. Add sound. We used standard PC speakers plugged into the wave shield. (Not shown.)

PROJECT 2

SUDO MAKE ME A SANDWICH ROBOT

This machine, inspired by a popular comic, relieves a pressing need for automatically prepared toasted cheese sandwiches. Motors dispense bread and cheese and insert the resulting sandwich into a toaster oven.

Credit: Adam Cecchetti

HACKERSPACE PROFILE: NYC RESISTOR

Location:

NYC Resistor

87 3rd Avenue

4th Floor

Brooklyn, NY 11217

www.nycresistor.com

Organizational type: LLC

Founding date: February 2008; moved to a new space in 2010

Number of members: 40

Dues: $75/month for members who teach 6+ classes per year, $115/month with no teaching requirement

Size of the space: 3,000 sq. ft.

Officers/Leaders:

- Raphael "Teuthis" Abrams, czar of membership
- Kellbot, czar of fundraising
- Herb "POTUS" Hoover, czar of classes
- Eric Skiff, czar of accounting

Notable equipment:

Laser cutter, vinyl plotter, large format printer, vacuum former, sewing machines, knitting machines, a MakerBot, a mini CNC mill, and a microscoe.

SPACE DETAILS

"We're on the 4th floor of an industrial building in downtown Brooklyn," said member Kellbot. "It's kinda grungy, but in a charming way. When we moved in the floor in the back room was so trashed we ended up building a plywood floor on top of it. The whole space is covered in laser cut crap and has a general 'crazy dude's garage' feel. There is no passenger elevator, but if there's a pressing need, the hackers can make it happen. "We do have the ability to, uh, enable the freight elevator should there be someone who can't take the stairs," she said.

Unlike many spaces, NYC Resistor doesn't have a formal organizational structure like a president and vice president. "We have 'czars' who are members who have volunteered to be the point of contact for some administrative tasks," Kellbot explained. "They're not really elected because we've never had more than one person vying for a position."

When a czar gets sick of the position, a new person rotates in.

Credit: Joe Curzon

▲ *The entrance to NYC Resistor.*

THE PROJECT

Participants: Adam Cecchetti and Bre Pettis

"We all know that life would be better if we could make a sandwich by typing 'Sudo Make Me a Sandwich' on the command line of a computer," wrote Bre Pettis in a blog post.

It all started with a cartoon. Nerd-favorite comic XKCD (xkcd.com) explores every facet of the geeky life from the perspective of wise-cracking stick figures. It "appears to have some sort of window into my head," Pettis wrote. "I swear I think of things and the next day, there is an XKCD comic about it."

This cartoon shows a lazy person asking his friend to make him a sandwich. The friend refuses but relents when the first guy repeats his request with *sudo* at the beginning. In Unix parlance, *sudo* is a command that overrides any other factor—the su stands for superuser, and do is the command. Cecchetti and Pettis played off the comic by creating a machine that prepares a toasted cheese sandwich when activated from the instruction of "sudo make me a sandwich" typed into a computer's command prompt. Working for 48 hours without sleep, the two hackers designed and built the machine in NYC Resistor's workshop.

Credit: Bre Pettis

▲ *Edith making a box at the hackerspace.*

Credit: Randall Munroe

▲ *The comic that inspired the creation of a sandwich-making robot.*

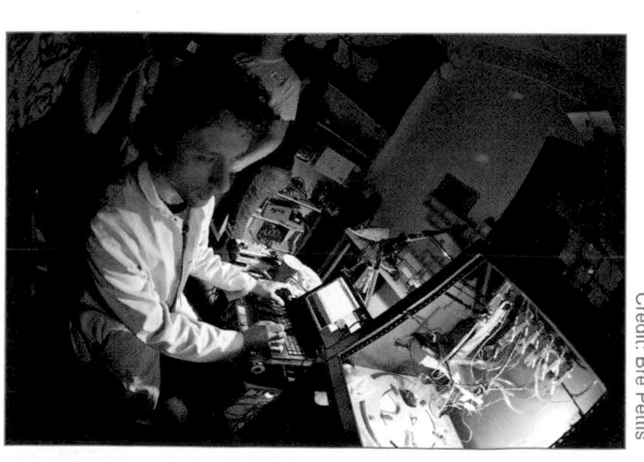

Credit: Bre Pettis

▲ *NYC Resistor member Raphael Abrams works the space's cocktail robot during a fundraising party.*

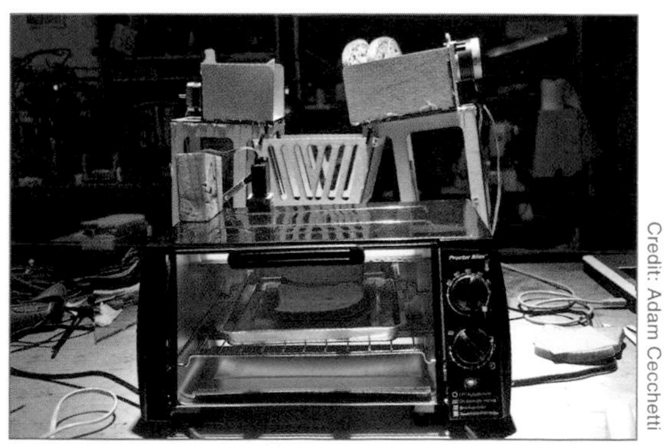

Credit: Adam Cecchetti

▲ *The Sudo Make Me a Sandwich Robot dispenses bread and cheese onto a tray, which is pulled into a toaster oven.*

The robot consists of a toaster oven with its back panel removed to allow a tray to be inserted. Above it, motorized bread and cheese delivery mechanisms are positioned, held up by a framework of laser-cut wooden panels that Pettis created in design program QCAD and output on NYC Resistor's laser cutter.

The dispensers work like the vending machines where the product is held in big spirals of wire. When the robot is activated, the bread spiral turns, dropping a slice of bread onto the tray. Then a similar spiral drops a slice of cheese onto the bread and, finally, another slice of bread is dispensed on top of the cheese. The tray holding the assembled sandwich slides into the toaster oven through its back panel and the sandwich is cooked. "The toaster oven needed a little modification," Pettis described. "A servo-controlled flap was put into place with some hinges to make it move slowly."

The robot's motors consist of two steppers, which control the cheese and bread systems, as well as two servos, which open and close the oven door and handle the sandwich tray. An Arduino manages the motors with the assistance of two RepRap stepper controllers.

Credit: Bre Pettis

▲ *Adam Cecchetti designs the robot's electronics; the circuit boards and motors will be installed in the wooden framework.*

Credit: Bre Pettis

▲ *The bread assembly works the same way as a snack machine—a stepper motor turns a spiral, which dispenses the slices one at a time.*

Credit: Adam Cecchetti

▲ *Pettis and Cecchetti "modified" the toaster oven with the help of a pair of bolt cutters.*

Credit: Bre Pettis

▲ *Team member Adam cuts a board on the space's band saw.*

Credit: Bre Pettis

▲ *The bread dispensing mechanism seen close-up. Note the way the spiral is connected to the motor: a glob of hot glue.*

Credit: Adam Cecchetti

▲ *An Arduino sits on a pile of design sketches.*

The final step was to program the whole assembly to trigger when someone typed the correct command into the interface. Of course, the command line isn't really necessary to start the robot. It could far more easily have been a simple button or other actuator; the typed command was simply following the spirit of the comic.

In any case, the command line interface isn't truly Unix—it would have been ludicrous to retask the "sudo make" command to send data via serial to the Arduino and thereby trigger the robot. Instead, Cecchetti and Pettis used a scripting language called Python to simulate the behavior of Unix. The script simply awaits two commands, either 'make me a robot' (which returns 'what, make it yourself!') or 'sudo make me a robot' (which sends the start command to the Arduino via a USB cable).

Another shortcut was that the oven doesn't turn on when instructed by the robot. Instead, when the operator wants to make sandwiches, the oven is turned on manually.

After a marathon hacking session, the robot was done and they were thrilled with the result. "This is one of those robots that I swear is alive," Pettis said. "The noises it made were like an animal, and it seemed that every time we looked the other way, it was coming to life and changing things with the setup."

KNOW YOUR MOTORS

While there are many flavors of electrical motors, most DIY projects involving motors use one or both of two basic types:

SERVO

A servo is a basic electric motor; run a current through the lead wire and it spins. Reverse the polarity and the motor turns backwards. Many servos also have a third wire that provides data to and from the microcontroller. The purpose of this data is to precisely control the speed and position of the motor. Most hobbyist servos have a built-in gearbox giving the user many options for speed and torque.

▲ *A HiTech hobbyist servo. Note the yellow data wire and the translucent enclosure revealing the gear assembly.*

STEPPER

By contrast, a stepper motor can divide its full rotation into a large number of steps. It has permanent magnets attached to it, interacting with a series of coils that create a magnetic field. As the coils are activated, the stepper turns to a certain position. The advantage is that the stepper can be precisely controlled by a microcontroller, and as a result, it is used for position-critical tasks like moving the imaging bar of a flatbed scanner. To do its job, a scanner needs to know where that imager is and be able to adjust its position to it down to the millimeter. Frequently, tinkerers employ specialized boards called *stepper drivers* to run their steppers.

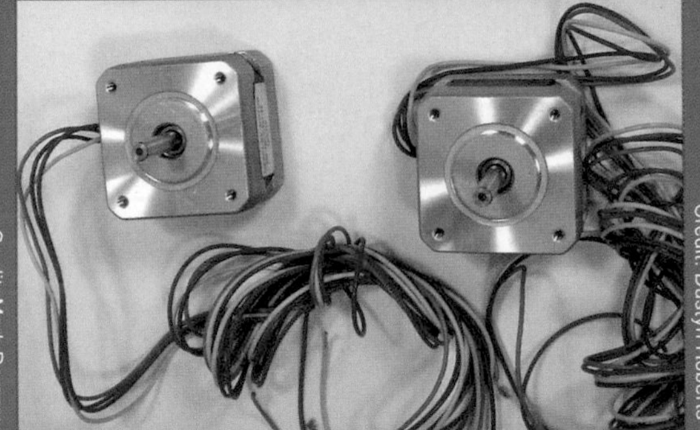

▲ *Stepper motors. Each stepper has four or more wires, making them much more complicated to run with a microcontroller.*

TOOL BIN

REPRAP STEPPER DRIVER

The RepRap stepper driver simplifies operation of a stepper motor. In addition to handling the motor's rotation, it accepts input from one or two limit switches that stop the motor if one of the switches is triggered, preventing potentially damaging collisions. The stepper driver is especially important for managing motors for an Arduino, which isn't suited for the task.

Credit: Zach Smith

▲ *Controlling a stepper—much less two or more—with a microcontroller can be a pain. Enter the stepper driver.*

▲ *A rear view of the robot shows how the tray and oven door are manipulated by servos.*

BUILD IT

Want to make your own? Cecchetti and Pettis provided a lot of details for creating your own Sudo Make Me a Sandwich Robot, with the Arduino code and support structure design available for download. Still, this isn't a shoo-in. You'll need to do a fair amount of tinkering to get it to work.

BASIC BUILDING STEPS

1. Download the chassis pattern from Thingiverse and cut it out: http://www.thingiverse.com/thing:332. The robot's chassis consists of thin sheets of plywood, laser-cut into the shape you see in the photos. You can download the DXFs (the format read by laser cutters) on Thingiverse.com. Alternatively, you can design and construct your own structure to support the dispensing mechanisms.

2. Construct the bread and cheese dispensing assemblies. These are essentially wooden boxes with spirals of stiff wire connected to stepper motors. Cecchetti and Pettis harvested the steppers from broken pieces of consumer electronics. You can probably use two random steppers with no trouble.

3. Remove the back of the toaster oven. This might be easy or hard depending on the oven you choose. The guys used a Proctor Silex 31117, but you can probably use any model as long as the back is removable. Just be sure you don't damage any of the oven's electrical components, including its heating elements.

4. Add servos. The two servos control the tray sliding mechanism as well as open and close the oven door. Cecchetti and Pettis used Towerpro MG995 servos. You can substitute other models, but make sure you choose ones with a lot of torque; the MG995 is rated for 10 kilos.
5. Wire up the Arduino and motors. You can connect the servos directly to the Arduino, but the steppers are more complicated and require RepRap Stepper Drivers to manage them. You can buy these from the MakerBot store: http://store.makerbot.com/electronics/assembled-electronics/stepper-driver-v2-3-fully-assembled.html. You can download the Arduino code from Cecchetti's site: http://shadowflux.com/sammich.c.
6. Add power. You can power your robot in any number of ways, but Cecchetti and Pettis used a regular computer power supply, which converts AC to DC.
7. Make sandwiches!

FURTHER READING

- Bre Pettis's blog entry: http://www.brepettis.com/blog/2009/2/27/sudo-make-me-a-sandwich-robot.html
- Bre Pettis's Flickr set: http://www.flickr.com/photos/bre/sets/72157614417997915/
- Adam Cecchetti's blog entry: http://adamcecc.blogspot.com/2009/02/sudo-make-me-sandwich-robot-brooklyn.html (which has a great video showing the project)
- Andrew Cechetti's Flickr set: http://www.flickr.com/photos/adamcecc/sets/72157614427262105/

PROJECT 3

NETWORKED GEIGER COUNTER

Members of Tokyo Hackerspace responded to the cataclysmic meltdown at the Fukushima nuclear power plant by hacking a Geiger counter to send its data to the Web.

HACKERSPACE PROFILE: TOKYO HACKERSPACE

Location:

Tokyo Hackerspace
5-11-11 Shirokanedai
Tokyo Japan
170-0052
http://tokyohackerspace.org/

Organization type: Sole proprietorship

Founding date: September 2009

Number of members: 22

Dues: ¥5,000/month or ¥50,000 annually

Size of the space: 170 sq. meters

Founders:

- Chris Shannon
- Akiba
- Karamoon
- Llhuga
- Lauren Shannon
- MRE
- Richard Frankum
- Dave Sonntag
- Sonicvis

Notable equipment: Milling machine, fully stocked kitchen equipment with restaurant-grade prep table and tools, library, white boards, basic tools

TOKYO HACKERSPACE DETAILS

"Our space is a nontraditional house with a great Japanese garden, a great kitchen, and outdoor BBQ area for communal cooking and eating," co-founder Akiba described. "We have three work tables, a couch, and various work areas. It is on a quiet side street not too far from train lines and the subway."

Credit: Mitch Altman

▲ *Immediately upon entering Tokyo Hackerspace, you realize that you're in a house—a Japanese house at that.*

Credit: Mitch Altman

▲ *Like every building in Tokyo, space is at a premium in the hackerspace.*

Credit: Mitch Altman

▲ *Hackerspace members and guests hang out in the social area. As one might surmise from this photo, most of the members are expatriate Westerners.*

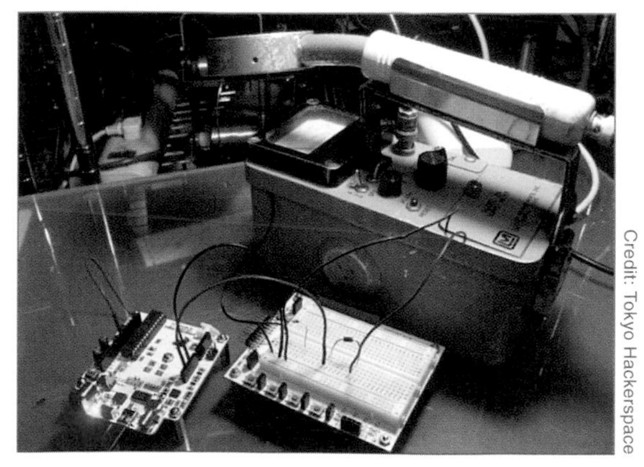

Credit: Tokyo Hackerspace

▲ *Tokyo Hackerspace, responding to the meltdown of a nearby nuclear power plant, connected an ancient Geiger counter to the Web via a wireless module.*

THE PROJECT

Participants: Akiba, Chris Shannon, Karamoon, Llhuga, Lauren Shannon, MRE, Richard Frankum, Dave Sonntag, and Sonicvis

On March 11, 2011, Japan was hit by a triple calamity: first, a 9-Richter earthquake, and then a tsunami. Both of these struck the nuclear power plant in Fukushima, resulting in the destruction of the facility and a massive radiation leak. "Riding the train to the hackerspace meeting, you could see the tension in the air," co-founder Akiba recalled. "Jaws were clenched, there was complete silence on the train, and everyone was glued to their phone screens to check for updates."

Added to the threat of radiation was the reality of power rationing. "You couldn't be sure if you could make it home at night because the trains could stop at any time," Akiba said. "We had to give ourselves a margin of two to three hours for any type of travel because the train schedules were so uncertain."

Food became scarce, power and communication channels were disrupted, and panicked tourists jammed onto planes to flee the country. The members of Tokyo Hackerspace, however, weren't panicking. In fact, they immediately began working on ways they could help. "A lot of the electronics guys went straight to work on putting together designs to scavenge power from the surroundings," Akiba said. "One really innovative design was to use car batteries as phone chargers. There were a huge amount of cars that were wrecked by the tsunami, and they were just lying around."

Credit: Tokyo Hackerspace

▲ *With power supplied by the Freakduino microcontroller board (described later in this project), the Geiger counter's alkaline batteries became a hindrance. The team insulated the batteries and attached leads from the external source, fooling the Geiger counter into thinking it was getting electricity from the batteries.*

Credit: Tokyo Hackerspace

▲ *The Freakduino Chibi interprets data from the Geiger counter and sends it wirelessly to the computer via another Freakduino.*

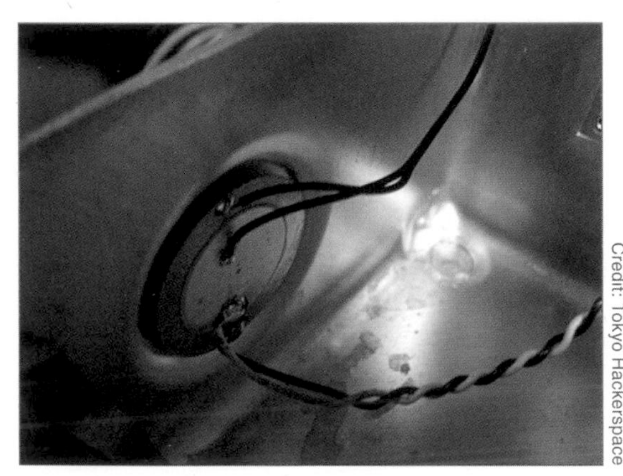

Credit: Tokyo Hackerspace

▲ *Jumpers attached to the Geiger counter's speaker send its analog signal to the microcontroller.*

One of the projects the group created was a cheap solar-powered "Kimono lantern" that could be left outside during the day and then used for illumination at night, during the frustrating blackouts that accompanied the disaster. The space built 150 of the lanterns using donations from other hackerspaces, and they're expecting as many as 500 more to be shipped to them from overseas.

While the rolling blackouts were an annoyance, the threat of radiation proved to be much more urgent. For a time after the disaster, the Japanese government refused to divulge radiation levels around the country. "The government would only release radiation readings once per day, and it would be disseminated to the public like a weather report," Akiba explained. "As you can imagine, people in the hackerspace were skeptical of government data." The fact that international organizations and foreign countries refused to believe the Japanese government's numbers didn't encourage the skeptics.

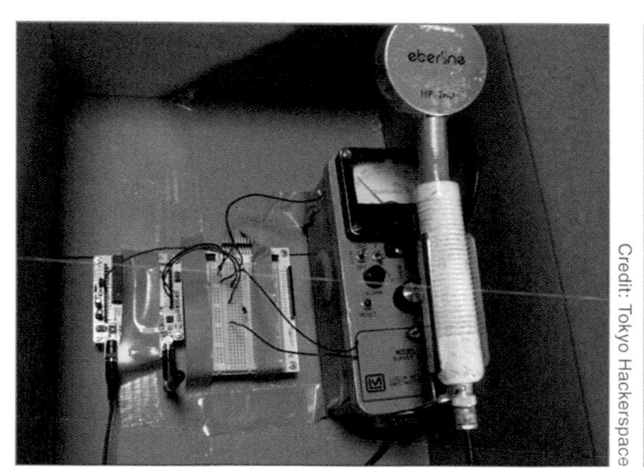

Credit: Tokyo Hackerspace

▲ *Akiba placed the Geiger counter and Freakduino in a cardboard box to protect them somewhat, while keeping them exposed to the air. The tape is to ward off damage caused by aftershocks.*

Credit: Tokyo Hackerspace

▲ *An oscilloscope displays the clicks of the Geiger counter, translated into data.*

"A day after the Fukushima explosions, we ordered 10 Geiger tubes off of eBay," Akiba said. "By that time, there was already a run on Geiger counters and most suppliers were sold out." About two days after the meltdown, Reuseum, a scientific surplus store located in Idaho, contacted Tokyo Hackerspace and offered to ship them a couple of analog Geiger counters. When the counters arrived, one was designated as a communal resource so members could bring it home to measure radiation levels where they might affect loved ones. As for the other counter, the team got to work figuring out how to connect it to the Web.

"Unfortunately, the Geiger counter was completely analog and there was no way to pull data from it," Akiba wrote in a blog post. "So, being the nerd that I am, I proceeded to hack it into what I wanted."

A Geiger counter is a device for detecting the presence and level of ionizing radiation. The critical component of the counter is the Geiger-Müller tube, an enclosure filled with an inert gas such as helium or neon. When ionizing radiation passes through the tube, it ionizes some of the gas, causing current to flow across a series of electrodes. Ordinarily, a Geiger counter emits clicks from a speaker corresponding with the radiation level it detects. When the radiation increases, the clicks become faster.

The team wasn't even sure if it was possible to digitize the clicks. "The first instinct was to probe the output, but it felt like it would be too risky," Akiba explained. The team eventually decided to try it and attached the speaker leads to the probes of a voltage meter. "It turns out that it was a single-ended supply from 0 to 8V. The voltage was slightly high, but if I limited the current, it should be okay for the Arduino." They wrote a Arduino sketch that looped until it detected a strong enough signal, indicating a click. They tested the feed and discovered that the Arduino was detecting every click. Success!

The next step was to connect the signal to the Web. The hackerspace was intrigued by the services offered by pachube.com, which displays real-time sensor feeds from sources around the globe. After the disaster, dozens of hobbyists and scientific institutions connected their devices to pachube. Most of these counters were located outdoors. "I found that the readings outside my apartment were about two to three times the readings inside," Akiba recalled. However, locating it outside meant that Akiba would have to run a data cable from the inside of his apartment to the balcony. "The temperatures were cold at night," he explained. "Since the radioactivity was mostly from airborne particles, I didn't want to have my door open all the time." Akiba made use of his own creation, a wireless data module called the Freakduino Chibi (see sidebar) to send the data to the computer remotely. With only a thin power cable supplying the setup with juice, Akiba's patio door could be closed.

Ultimately, the aggregated data from the various Geiger counters connected to pachube.com confirmed the Japanese government's numbers—but had an added value in that the sensor data was continually updated, potentially revealing radiation spikes that might otherwise go unreported with the government's once-a-day approach. Perhaps more importantly, it served as a test case for a hackerspace's ability to mobilize their members' collective skills and motivation to help the community in a time of need.

AKIHABARA

Credit: Ivva (Community Commons)

▲ *Tokyo's Akihabara is a tinkerer's haven, with store after store stuffed with gadgets, tools, and electronic components.*

Credit: Ivva (Community Commons)

▲ *This display of multimeters shows the dizzying variety that is part of Akihabara's allure. Not content to offer a handful of options, most shops in the district give the shopper as many choices as possible.*

Akiba, the hacker who spearheaded the Networked Geiger Counter project, takes his handle from the nickname of Akihabara Electric Town, a commercial district in Tokyo known for its electronic components, computers, and comics. "Basically, it's the jewel of Tokyo if you're a tech geek," Akiba described. "Just imagine that you're a chef, you live right next to one of the biggest, most expansive farmers' markets in the world, and what kind of influence that would have on you. The fact that you can see, touch, and browse something similar to the Digikey catalog is unexplainable."

Akiba—the district, not the hacker—is packed full of tiny, crowded shops featuring shelves of gizmos, robots, comics, and toys. Its specialty, however, is electronic components—every part a person could ever hope for. An electronic hobbyist visiting the place typically announces that he is in heaven, poring over rack after rack of brightly colored electronic switches and LED enclosures, mysterious connectors, bright chrome toggle switches, spycams, soldering equipment, magnets, pulleys, gears, and gaskets.

"Visiting the shops, especially the surplus shops, was also extremely educational to me," Akiba said. "There were so many parts that I've never seen before, like industrial controls, specialty tubes and valves, motors, pumps, sensors, etc. It took me about five years of digging around in each shop and sniffing around nondescript buildings to build up my current knowledge of the area."

Akiba and a friend recorded a series of videos giving an overview of Akihabara. You can find them here: http://www.youtube.com/user/wipinmtl

Update: As we went to press with this book, Akiba reported that he had placed his Geiger counter in a cool case, making transportation easier.

Credit: Iva (Community Commons)

▲ *Not called the Electric Town for nothing, Akihabara's shops offer all manner of electrical devices ranging from the latest robots to these pedestrian-seeming fluorescent light fixtures.*

▲ *Akiba's Geiger counter is now much more portable.*

TOOL BIN

FREAKDUINO CHIBI

Akiba created the Freakduino Chibi to simplify setting up and configuring wireless data transfers. At its heart, it's an Arduino with wireless capability. "ZigBee is complicated to set up," he said, referring to a competing protocol that governs wireless networking. He even found the set up of the XBee (http://www.digi.com/xbee/), a hobbyist-friendly wireless board, to be excessively complicated. "I just wanted something that allowed people to send and receive data, and I focused on keeping the main API as small as possible." The API is the programming interface, and Akiba limited the number of commands to three total, to keep it simplified.

In essence, any data connection that would travel over a cable can be effected via radio waves using a module like the Freakduino to interpret the traffic. By combining the wireless module with the functional equivalent to an Arduino board, Akiba added to its ease of use—thousands of hobbyists know and use Arduinos, and countless Arduino-focused tools and libraries already exist.

Akiba was concerned not only with creating a simplified and powerful product, but also one that wouldn't be too expensive. "It had to be cheap enough that it wouldn't be considered a big expense to people that just wanted to play around with Arduinos and wireless communications," he said. "That's why I made it available in kit form as well as in assembled and enclosure-based configurations."

Credit: Akiba

▲ *The Freakduino Chibi packs all the features of a standard Arduino and adds wireless connectivity to the mix.*

If you're interested in buying one of these boards, you can find them on Akiba's online store: http://www.freaklabsstore.com/.

TOOL BIN

PROCESSING

Processing, the programming language Tokyo Hackerspace used to connect to pachube.com, is an open-source language designed by Ceasey Reas and Benjamin Fry while they were at MIT Media Lab's Aesthetics and Computation Group. Though based on Java, a robust programming environment favored by professionals, Processing was designed to make computer programming accessible to artists, hobbyists, and others without the inclination or ability to learn more advanced programming skills. It permits a casual user to create visual displays very rapidly, without the need for a compiler. This artist focus is so ingrained in the platform that its programs are actually called *sketches*.

Similarly, electronics hobbyists have come to love Processing's easy interface, and the language has been adapted to form the basis for Arduino, the language that runs Arduino boards. Because of this relationship, Arduinos interact very well with Processing sketches.

Credit: Peter Kirn, Community Commons

▲ *Peter Kirn's Processing sketch creates a visualization based on an audio feed.*

BUILD IT

While hopefully you won't have a need to monitor the local radiation levels, Tokyo Hackerspace's networked Geiger counter will do the trick. Here's how to do it:

1. Solder jumpers to speaker. Remove the Geiger counter's case and solder on jumpers to the speaker. The positive lead goes to an analog port of the Freakduino, and the ground wire connects to the ground port.
2. Connect the counter's battery leads to the Freakduino's 3.3V port. Tokyo Hackerspace did this by removing the Geiger counter's ancient batteries, insulating them so they aren't exposed to voltage, then taping wires onto them to make the Geiger function as if it was running off of battery power.
3. Install the code on Freakduino. You can download Tokyo Hackerspace's code here: http://freaklabs.org/images/stories/2011-03-22%20Geiger/Arduino_geiger.zip.
4. Connect the second Freakduino to a desktop PC hooked up to the Internet. Load Processing onto the PC and run the following sketch: http://freaklabs.org/images/stories/2011-03-22%20Geiger/Processing_geiger.zip.
5. Set up a pachube account. Go to pachube, set up your account with the service, and then publicize your feed!

FURTHER READING

Akiba's post about the Geiger counter project:

http://freaklabs.org/index.php/Blog/Misc/Hacking-a-Geiger-Counter-in-Nuclear-Tokyo.html

PROJECT 4

GLASS BLOCK LED MATRIX

The members of Cincinnati hackerspace Hive13 wanted to dress up a boring glass block wall. What better way than an interactive, multi-colored light-emitting diodes (LED) matrix?

HACKERSPACE PROFILE: HIVE13

Location:

Hive13

2929 Spring Grove Avenue

Cincinnati, OH 45225

www.hive13.org

Organizational type:

Ohio nonprofit, federally recognized 501(c)(3) scientific and educational charity

Founding date: July 2009

Number of members: 21

Dues: $100 Cornerstone, $50 Basic, and $13.37 l33t (elite).

Size of the space: 3,500 sq. ft.

Officers/Leaders:

- Dave Menninger, president
- Craig "zombieCraig" Smith, treasurer
- Chris Anderson, secretary, CTO
- Chris Davis, COO
- Jason "fuzz" Bailey, board member
- Mike Linville, board member

Notable equipment: MakerBot, drill press, table saw, oscilloscope, silhouette paper cutter, large vinyl cutter, full kitchen, projectors, and lathes

SPACE DETAILS

"Our space is in a building that has seen many uses over the years," Hive13 president Dave Menninger described. "Within the same building our neighbors are an artist's apartment and his art studio, a stained glass maker, a paper company, an arcade machine company, students' apartments, and a fitness/boxing club. We have a full kitchen, an enclosed 'office' area, and a larger open area. We have recently set up a huge projection screen made from a painter's drop cloth."

Credit: Chris Davis

▲ *The main work area at Hive13.*

Credit: Paul Vincent

▲ *A class on reverse engineering draws a full house at the hackerspace.*

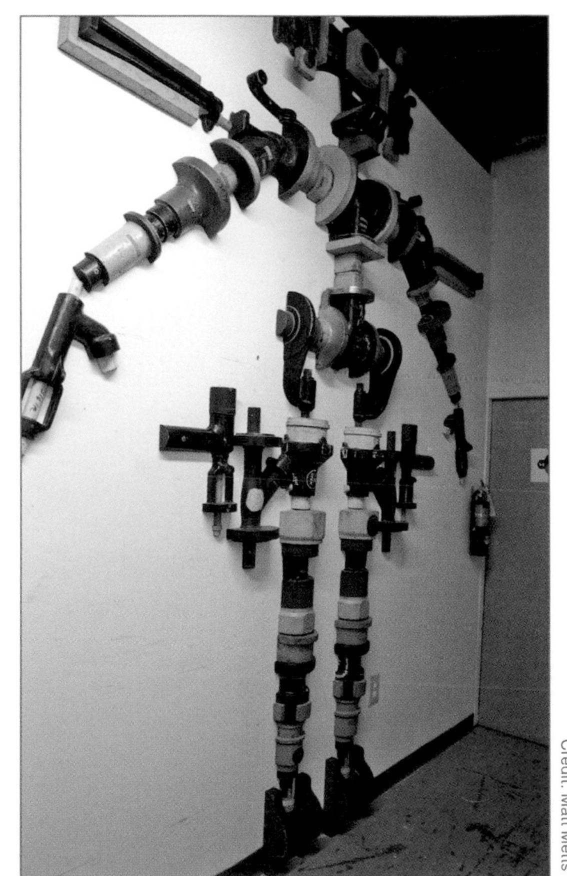
Credit: Matt Metts

▲ *This sculpture by Hive13 member Rob Fronk was built using salvaged industrial equipment.*

THE PROJECT

Participants: Nathan "Deckmaster" Curde, Chris Davis, Dave Menninger, Jon Neal, James Stoneburner, and Paul Vincent.

Credit: Paul Vincent

▲ *The exterior of Hive13's building shows the LED matrix in "plasma" mode.*

When the hackers of Hive13 moved into their new space, they spotted the five-foot glass block window and immediately thought to make it into an LED matrix. This is a series of LEDs arranged in a grid like the pixels of a computer display. With each light able to accept commands independently of the others, all sorts of patterns, animations, and text displays can be generated. Even better, when the matrix is laid out on top of a glass block wall, people passing by outside the building get to experience the display of lights as well.

The matrix took about a year to complete. Like many hackerspace projects, the idea came early, but the work took a while to gestate as components were purchased and individual members' interest peaked and faded. Sometimes, too, there were false starts. "Part of the reason for the long duration for the project is after we had put significant effort into the original design, we radically changed our implementation," project lead and Hive13 president Dave Menninger described. The original design called for RGB LEDs, a microcontroller called a *rainbowduino* optimized for running large numbers of LEDs, Cat 5 ethernet cable for the wiring, a framework of PVC pipes to support the LEDs, and prototyping boards to hold components.

"We ran into a long series of technical obstacles," Menninger described. "Back current, induction, un-grounded pulse width modulation, 600 solder joints." After slogging through all of these obstacles, the participants realized they could build a better version with much less effort.

After a while the group settled on the new design: Instead of RGB LEDs, the newer matrix utilized ShiftBrites, which are high-powered RGB LEDs with advanced features such as over-temperature control and an integrated voltage regulator. Instead of a rainbowduino, they used a regular Arduino and a ShiftBrite shield; the latter is an add-on to the Arduino for controlling the ShiftBrites. They built a series of pine slats to replace the PVC piping.

"After settling on the new design and getting all the parts, construction and coding went very quickly—on the scale of hours, not weeks," Menninger said. The ShiftBrites were strung together in a strand of 56, with one LED for each glass block. The ShiftBrites were secured with ordinary binder clips from an office supply store, allowing the LEDs to easily be reconfigured.

Coming up with the money to buy the parts was a project in itself. In most hackerspace projects, participants provide the funding—often the organization lacks extra money for this sort of thing. To help scrape together the funds to buy all the parts, Menninger asked for donations via the fundraising site Pledgie and was able to collect all but about $60 of what the project required.

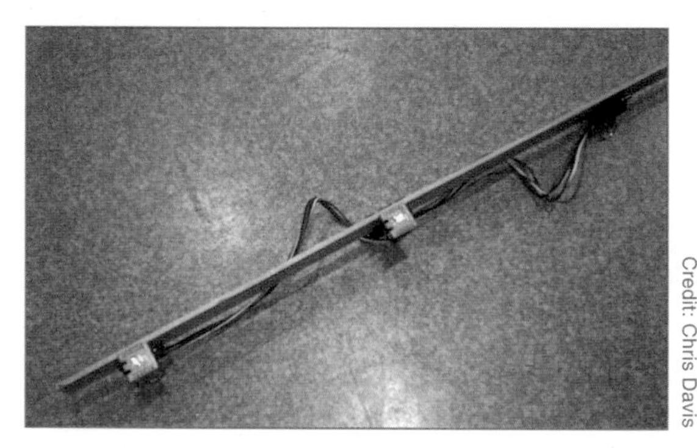

Credit: Chris Davis

▲ *ShiftBrites are attached to wooden slats with office binder clips. Need to adjust the locations of the LEDs? No problem.*

Credit: Paul Vincent

▲ *The ShiftBrites are arranged in a daisychain that covers all 56 blocks of Hive13's window.*

Credit: Paul Vincent

▲ *Chris Davis programs the Arduino board that controls the LEDs.*

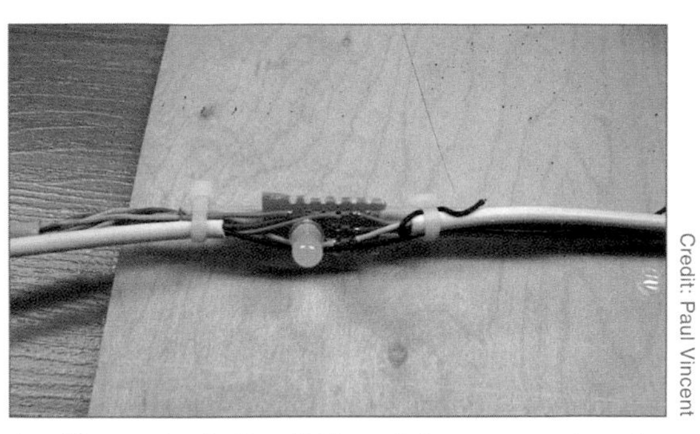
Credit: Paul Vincent

▲ *The project's first LED modules were painstakingly hand-soldered. Despite the custom design, the creators still weren't happy.*

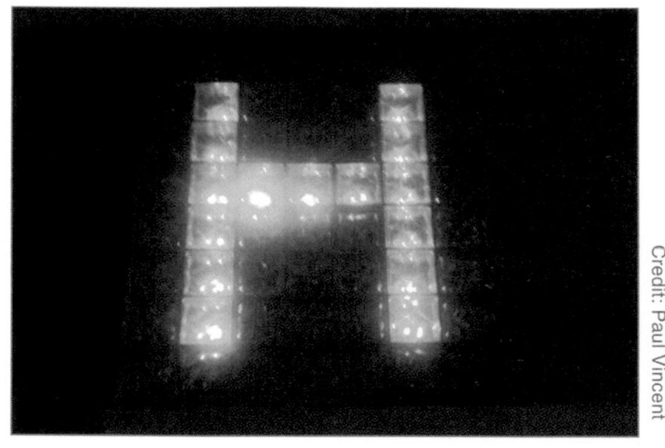
Credit: Paul Vincent

▲ *The matrix can show letters, too. This flashing H advertises the hackerspace to passers-by.*

Credit: Paul Vincent

▲ *The team tested the modules' output after adopting ShiftBrites in place of regular LEDs.*

Credit: Paul Vincent

▲ *The original framework was built from PVC pipes, but this was abandoned in favor of a simpler solution.*

With the project complete, Hive13 has a huge light-up display that can be used for outreach or decoration. It has been programmed for different modes, ranging from an animated "plasma" display that shifts through a series of colors to simply a series of flashing letters spelling out HIVE to identify the space. It is, in essence, a blank "screen" that can be used for a variety of purposes.

THE MIDWEST HACKERSPACE TOUR

In 2010, Noisebridge founder Mitch Altman set out with Boston hacker Jimmie Rodgers as well as *MAKE Magazine* writer Matt Metts. The goal was to visit as many Midwest hackerspaces as they could.

Beginning in Detroit, the group made a quick trip north to Ontario and visited four spaces in that province. Returning to the United States, they drove through Michigan, Kentucky, Ohio, Indiana, Missouri, Iowa, Illinois, and Wisconsin, holding workshops along the way. All told, Altman and his friends traveled 3,573 miles and visited 18 U.S. and Canadian hackerspaces in 26 days, giving talks and workshops at each stop. The group, along with a corps of volunteers, taught more than 1,500 people how to solder.

"The best part for me was meeting all the incredible people who create hackerspaces, make them happen, and keep them happening," Altman said. "These are some of the most creative, intelligent, and fun people in the world. The hackerspace movement is a great filter for humanity."

For their part, the Hive13 members were excited to play host for Altman and his crew. "Having Mitch and company visit our space was awesome in several ways," hackerspace president Dave Menninger said. "For one thing, it got a lot of people in the door that were new. Another good aspect of Mitch's visit was that it prompted us to acquire a big set of soldering irons, wire snips, etc., so that we are now outfitted to have classes on soldering again with a big group of people. Mitch kept saying, 'You don't need me. You can do this again.' So we got some confidence in ourselves from that."

Credit: Matt Metts

▲ *Hive13 members gather around Mitch Altman (seated, with wrist bands) as he teaches a class on soldering.*

TOOL BIN

LIGHT-EMITTING DIODES

One of the components that hobbyists reach for the most is the LED, essentially the light bulb used in electronics projects. They offer many advantages over incandescent bulbs, including a longer life, smaller size, lower energy consumption, and greater durability. They also feature a greater range of wavelengths, ranging from infrared to ultraviolet, and certain types of LEDs (known as RGB LEDs) are also capable of changing their color to any hue in the visible spectrum, though this requires the help of a more complicated electronic circuit.

Credit: Matt Metts

▲ *A matrix of LEDs awaits commands from a microcontroller.*

BUILD IT

The glass-block window at Hive13 called out to the organization's members, who resolved to build an LED matrix to turn the window into a light-up display. If you have a window like this, or simply want to create your own programmable LED matrix, follow these directions.

Here's how the project comes together:

1. Determine how many LEDs you'll want. The advantage of a glass-block window is that the LEDs' light will be diffused: that is, the whole brick will seem to light up. These instructions assume that you'll be working with a 7×8 grid, or 56 LEDs.
2. Assemble the wooden framework. Hive13 used slats of pine, 5/8"×70" long, nailing them in place over the rows of glass bricks.
3. Wire up the ShiftBrites. You'll want them in a single long string, sort of like a string of Christmas lights. Arrange them along the wooden slats and secure each ShiftBrite over its respective glass brick with a binder clip.
4. Connect the Arduino and its ShiftBrite shield. You can buy ShiftBrites, a ShiftBrite Shield, and cabling from macetech.com/store.
5. Plug in your power supply. The number of ShiftBrites you want to power determines what supply you use; Hive13 used a 9-volt, 4.5-amp LED switching power supply. The power and ground cables plug into the first module and are daisy-chained to all the others.
6. Program the Arduino. The matrix is a blank slate, and you can create whatever designs your imagination—and coding skills—allows. If you need somewhere to start, you can download Hive13's Arduino sketches from https://github.com/Hive13/windowmatrix.

PROJECT 5

BRONZE-MELTING BLAST FURNACE

It's traditional for hackerspaces to create a coin design with the hackerspace logo and then output it on a 3D printer. Quad Cities Co-Lab took their coin design one huge step forward by making theirs in bronze with the help of a homemade blast furnace.

Credit: Zak Kupua

HACKERSPACE PROFILE: QUAD CITY CO-LAB

Location:

Quad City Co-Lab
1033 East 53rd Street
Davenport, IA 52807
www.qccolab.com

Organizational type:

Iowa nonprofit, 501(3)c pending

Founding date: April 2010

Number of members: 20

Dues: $30/month

Officers/Leaders:

- Steve Hamer, president
- Jeremy Borchert, vice president
- Ben Ziegler, secretary
- Arron Lorenz, treasurer

Size of the space: 4,200 sq. ft.

Notable equipment: Welder, blast furnace, drill press, table saw, and MakerBot

Space details:

Former commercial space divided into three major rooms to separate louder/dustier activities from the rest. There's a conference room, classroom, studio, server room, and kitchen. The organization removed walls and carpet in the shop area and reworked the electrical and network.

Credit: Zak Kapua

▲ *The entrance to QC Co-Labs. Unlike many hackerspaces, the space wasn't originally industrial in nature, but rather a standard office.*

Credit: Matt Metts

▲ *A soldering class at QC Co-Lab has members assembling electronic kits.*

Credit: Zak Kapua

▲ *The space's electronic tinkering workbench, already in place even as the organization moved into its new facility.*

THE PROJECT

Participants: Dave Hinkle, Steve Hamer, and Chris Cooper, along with various others

Credit: Zak Kapua

▲ *The furnace assembly, ready for action.*

Credit: Zak Kapua

▲ *Team members gather in QC Co-Lab's workshop to plan their next moves.*

"Our first purchase as a hackerspace was the CupCake CNC," Quad Cities Co-Lab founding member Dave Hinkle recalled. This machine, more properly known as a MakerBot CupCake CNC, is a 3D printer that forms three-dimensional objects out of strands of melted plastic. Once their MakerBot arrived, Hinkle and the other members spent the next two days assembling the machine and the rest of the week fine-tuning it.

"The guys told me that it was becoming traditional for a hackerspace to do a coin design on the CupCake to represent their space," Hinkle said. "I thought, what could be more fun than taking the idea one step further and casting the coin in bronze?" Hinkle created a coin design in Blender using QC Co-Lab's light bulb logo.

"Basic casting in bronze and aluminum is cheap, fun, and very showy," Hinkle said. "I knew it would make a great demonstration for our grand opening. We only had a week to go, and there was no time to purchase a commercial furnace and have it shipped to us, so I decided to go the DIY route."

Quick research showed that a blast furnace is actually pretty simple: a fuel source, lots of oxygen to help the fuel burn, and insulation to concentrate the heat. The design they chose for the grand opening consisted of a charcoal-fueled furnace and a steel crucible, but they later refined the project as follows: the furnace is made of a metal trash can lined with refractory cement—a mixture of Portland cement, sand, perlite, and fire clay. Oxygen is pumped in with the help of a repurposed dog-grooming hair dryer, and the fuel is a tank of liquid propane. "The propane is fed through a weed burner I bought at Home Depot in the welding section," Hinkle explained. Heated inside the furnace is a ceramic crucible, and when it's ready, the molten metal is poured out of the crucible into the mold.

"These furnaces will melt aluminum very easily," Hinkle said. "Bronze has a much hotter melting temperature, so it's somewhat harder." QC Co-Lab's propane-fueled furnace goes from cold to melting bronze in 7 minutes.

The night of the open house, everything was ready. "There were hundreds of people filing through there throughout the day," Hinkle recalled. "It felt like a busy expo center." A crew filming a documentary interviewed participants while the Co-Lab's members gave demonstrations and tried to print the coin shape. Hinkle and friends created the mold using a wooden frame filled with green sand, a fine-grade silica mixed with bentonite. Then the plastic blank from the MakerBot was pressed into the sand to form the pattern.

"Things started to quiet down as it got dark," Hinkle said. "Ben Ziegler did an awesome display with his Tesla coil, and I fired up the furnace." As the furnace melted the bronze, Hinkle turned off the blower and gingerly removed the crucible. "The molten metal is unreal in person," he recalled. "It flows like engine oil, but it glows bright and warm. It feels like you're pouring thick, viscous liquid sunlight from a cup."

Credit: Zak Kapua

▲ *The liquid bronze flows out of the crucible, igniting the wooden frame of the mold.*

Credit: Zak Kapua

▲ *The MakerBot-produced blank (the white disk) surrounded by four bronze castings. The one at the top still has its sprue attached.*

Credit: Zak Kapua

▲ *Molten bronze, and the refractory concrete surrounding the crucible, glow orange.*

Credit: Zak Kapua

▲ *Visitors cluster around as Dave Hinkle (in the leather apron) monitors the furnace.*

Credit: Zak Kapua

▲ *Hinkle gingerly extracts the crucible and prepares to pour.*

After the poured metal cooled, the team broke open the molds to inspect their work. Success! While there were some rough edges and pieces that needed to be removed, the coins cleaned up beautifully.

"This project never would have happened without the guys at the hackerspace pitching in," Hinkle said. "It really has felt like a life-changing event for me, to suddenly have a community of people to work on projects with. The opening of the QC Co-Lab has been an amazing experience."

SAND CASTING TERMS

Metal casting artists have their own terminology. The following are some commonplace terms used in sand casting:

blank: The object to be duplicated in the mold.

cope: The top of the mold.

crucible: A heat-resistant vessel used to hold molten metal.

drag: The bottom of the mold.

dross: Excess slag at the top of the molten metal.

furnace: Used to heat up the metal to melting temperatures.

green sand: Wet sand made up of silica sand, bentonite, and anthracite, it's used to make the mold's shape.

ingot: Lumps of metal that are melted in the crucible.

parting dust: Fine, moisture-repelling dust used to keep the metal from sticking to the pattern.

pattern: The indentation in the sand in the shape of whatever you're molding.

refractory: A substance that retains its strength at high temperatures.

sprue: The channel through which the molten metal travels into the mold. After the molding process is complete, the sprue typically must be removed from the casting.

TOOL BIN

MAKERBOT 3D PRINTER

Most hackerspaces have a MakerBot, a 3D printer designed by MakerBot Industries in New York. MakerBot CupCakes and Thing-O-Matics are sold for $750 and up, making them some of the most affordable 3D printers around, and therefore putting them in the sights of hackerspaces, which are usually just scraping by. It doesn't hurt that the machines were themselves designed in a hackerspace. MakerBot Industries was founded by Adam Mayer, Bre Pettis, and Zach Smith of NYC Resistor. The connection is so strong that the business and the hackerspace are next door to each other.

A MakerBot works by melting a long strand of plastic filament and laying it down in precise patterns, building up a 3D object layer by layer. The resulting prints aren't as strong or attractive as custom-molded parts, but they are quicker and cheaper to produce.

▲ *QC Co-Lab's MakerBot CupCake with a homebrewed filament spindle behind it.*

Credit: Zak Kapua

BUILD IT

The hackers at QC Co-Labs wanted to make a big impact in their grand opening. Converting a digital design to a 3D bronze object totally fit the bill. Want to do the same?

Here's how the project comes together:

1. Create your blank, the item to be duplicated. Design your blank in a 3D modeling program, such as Blender (blender.org), and output it on your 3D printer. Alternatively, find an object on thingiverse.com and print it.
2. Build the mold. This is a complicated process that is thoroughly described on the sand casting website www.foundry101.com.
3. Assemble the furnace. Cut a hole in the side of the trashcan and line the inside with refractory cement.
4. Attach the burner. Hinkle used a propane weed burner from a home improvement store, and this controls the flow of propane into the furnace, sort of like the burner on a gas grill.
5. Add the blower. This supplies oxygen to the flame, allowing it to burn hot enough to melt metal.
6. When you're ready to fire up the furnace, add the crucible to the furnace and fill it with metal ingots.
7. When it has melted, pour the metal into the mold.
8. Let it cool for 15 minutes, and you're done!

PROJECT 6

MILKYMIST VJ CONSOLE

After years of using ordinary PCs for interactive VJing, the team at Paris hackerspace /tmp/lab decided to build their own, creating Milkymist, a custom board that generates visualizations based on sensor input and musical rhythms.

Credit: Sébastien Bourdeauducq

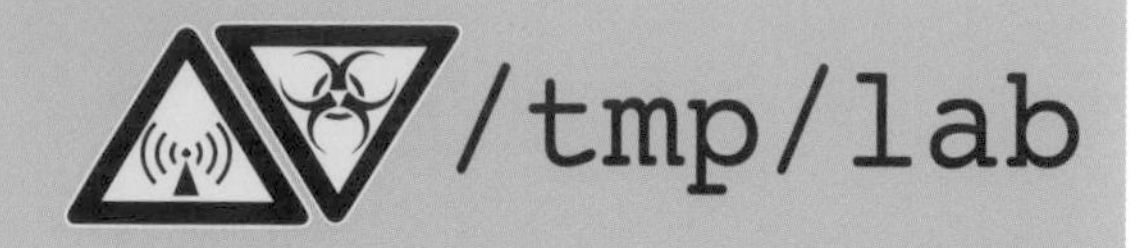

HACKERSPACE PROFILE: /TMP/LAB

Location:

/tmp/lab

6bis rue Lé]on Geffroy

94400 Vitry-sur-Seine

Paris, France

http://www.tmplab.org/

Organizational type: Nonprofit

Founding date: September 2007

Number of members: 30

Dues: 30€/year

Size of the space: 100 square meters

Officers/Leaders:

- Stephanie "Ursula", president
- Dermiste, treasurer

Notable equipment:

Ikos Pegasus ASIC emulator, chemicals (for chip decapsulation), microscope, DIY biodiesel reactor, kiln, shower, washing machine, RepRap, OLPCs, and a Siemens BS11 base transceiver station

SPACE DETAILS

"Rent is hard in Paris, expensive," co-founder Sébastien Bourdeauducq described. "We got an offer from an artist collective to have an artist space outside of Paris in the industrial suburb of Vitry-sur-Seine. The building owner temporarily gave us the space for no rent." The space's name comes from the temporary nature of this arrangement.

"The goal was to provide an infrastructure first," Bourdeauducq explained, "and let 1,000 beautiful projects blossom in this fertile environment: open Source, hardware, cultural and artistic events, activism, etc. We wanted everyone to see the /tmp/lab and say 'Oh... it's simple, let's build one with my friends in my town.'"

Credit: Paula Vélez

▲ *The door to /tmp/lab lays out members' politics for visitors.*

Credit: Paula Vélez

▲ *A /tmp/lab member hacks in the main work area.*

Credit: Girard Alexandre

▲ *Comfort in the tight confines of /tmp/lab can be a rare commodity.*

Credit: Paula Vélez

▲ *A sign at /tmp/lab touts Phack (phack.fr), an organization that supports Parisian hackerspaces.*

Credit: Sébastien Bourdeauducq (CC-BY-SA)

▲ *A beauty shot of the finished Milkymist One, showing off its laser-cut case.*

THE PROJECT

Participants: Sébastien Bourdeauducq, Yann Sionneau, Joachim Steiger, Lars-Peter Clausen, Takeshi Matsuya, Wolfgang Spraul, Adam Wang, and Michael Walle.

Sébastien Bourdeauducq of /tmp/lab used to do interactive VJing, where a club deejay uses a PC to generate video effects to go along with the music, with patterns and colors generated by the audio like a music program's visualizer. He discovered that he really didn't like using PCs—they were heavy, took a long time to set up, and if they lost power everyone in the club would watch as the computer painstakingly rebooted. Bourdeauducq wanted a dedicated system that could run his whole show from a small box.

▲ *Milkymist being demonstrated; the VJ runs music off his laptop while Milkymist handles the visualization.*

While Bourdeauducq was already very technically savvy—he ultimately used Milkymist for his master's thesis—he still needed help with the project and began looking for partners. "Geeks often disregard this project because they think it is too expensive and/or too complex," he explained. "It is exacerbated by the fact that DIY blogs adopt a similar attitude and largely overlook the project, making it even harder to reach out for people."

Ultimately, he found a team by directly approaching people he thought might be interested and by giving presentations at hacker conventions such as the CCC Congress and Notacon. They come from all over—for instance, Joachim Steiger hails from the Raumfahrtagentur hackerspace in Berlin and Lars-Peter Clausen is a member of Chaos Computer Club Hamburg. Bourdeauducq, along with Spraul and Wang, focused on laying out and producing the Milkymist board, while Sionneau, Clausen, Matsuya, and Walle developed the software. Meanwhile, Steiger designed and laser-cut the case.

The greatest challenge for the design team was making the Milkymist as small and handy as possible. That meant going with a system-on-a-chip (SoC) instead of merely using a PC. A SoC is essentially a big microcontroller, a low-powered computer with all the components like the microprocessor, memory, power management circuits, display controller, and so on integrated into a single chip.

"On the surface, Milkymist is promoted as a visual synthesizer," Bourdeauducq said. "But

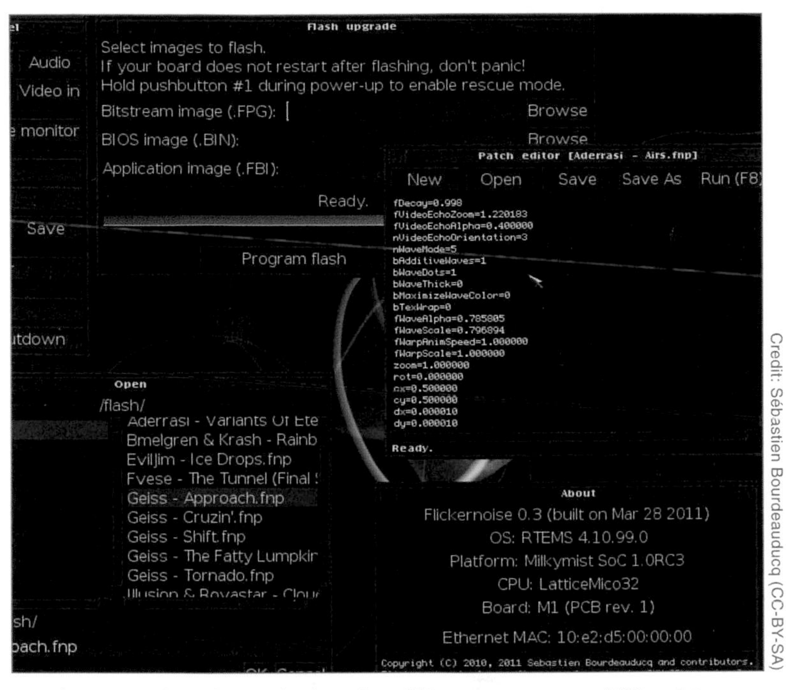

▲ *A screen shot shows the interface VJs use to operate Milkymist.*

Credit: Sébastien Bourdeauducq (CC-BY-SA)

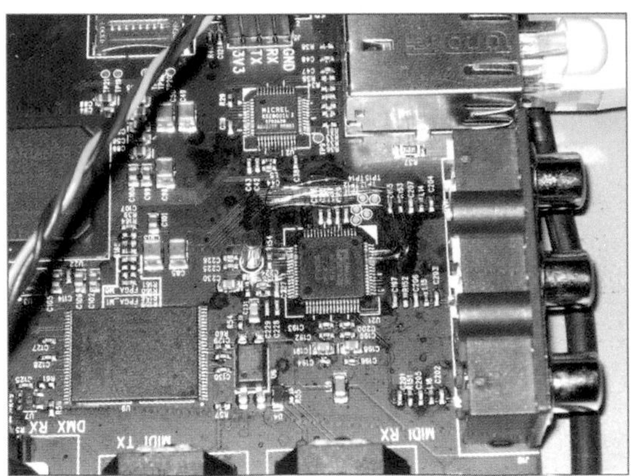

▲ *The tininess of Milkymist's surface-mount components (see Project 16, "LED Matrix Gaming System") made testing wiring changes a particular challenge.*

Credit: Michael Walle

it is also the leading open source system-on-chip design. It is today the fastest open source system-on-chip capable of running Linux, and it comes with an extensive set of features and graphics accelerators."

By using an FPGA for the central chip, advanced users can modify the design, either to customize the product for their own use or to contribute to the open source design. "This makes Milkymist the platform of choice for both anyone looking to build or use a fast open source processor and for the mobile VJ," Bourdeauducq said. FPGA, which stands for field programmable gate array, is the physical chip upon which the SoC is programmed. Unlike many integrated circuits that are configured in the factory, an FPGA can be reprogrammed in the field, hence the name. It is, nevertheless, the realm of the advanced programmer and probably too difficult for a neophyte to configure without a lot of study.

Credit: Sébastien Bourdeauducq (CC-BY-SA)

▲ *Milkymist's RAM presented a consistent problem early on and required a great deal of tinkering to get it working the way the team wanted.*

Credit: Takeshi Matsuya

▲ *The Milkymist board runs a version of the GNU-Linux operating system.*

After designing the board, the team set about manufacturing it. Because of the space constraints of the small system, it made more sense to use surface-mount components, which are much smaller and solder directly to the PCB, rather than through-hole components, which attach by wire leads threaded through holes in the board. The downside to SMD is that the manufacture can be quite difficult to do by hand.

Bourdeauducq traveled to the Minbo electronics factory in Taiwan and observed the entire manufacturing process. The six-layer PCBs have solder paste silk-screened onto them; then a manufacturing machine called a pick-and-place adds the SMD components. Next, the whole thing goes through a reflow oven to melt the solder. After the assembly, boards go through an automated optical inspection machine to detect soldering problems early. If the boards pass inspection, they are flashed with their initial architecture. Finally, self-test software is run on them to verify that all peripherals work as expected.

For VJing software, the team created Flickernoise, an easy-to-use interface that allows for the creation of visuals that react to the music with the help of control interfaces like MIDI (digital music) signals, video-in, DMX512 (digital stage lighting controllers), and even infrared remote controls. Flickernoise, like Milkymist, is open source and can be downloaded for free.

Unlike the software, however, the Milkymist team can't be given away for free. While Bourdeauducq initially intended Milkymist for his own use, it gradually became apparent that the team might have a commercial product on its hands. "As the technical problems gradually went away, I wanted to see 'how deep the rabbit hole goes' and make a fully fledged product out of it," Bourdeauducq recalled. "The project took a definitive turn in December 2009, when I met with Wolfgang Spraul, a former Openmoko employee, who offered to fund and take care of the production. This sped things up a lot."

The team sold about 35 Milkymists in the first four months, marketing the product as an "early developer kit" for hackers.

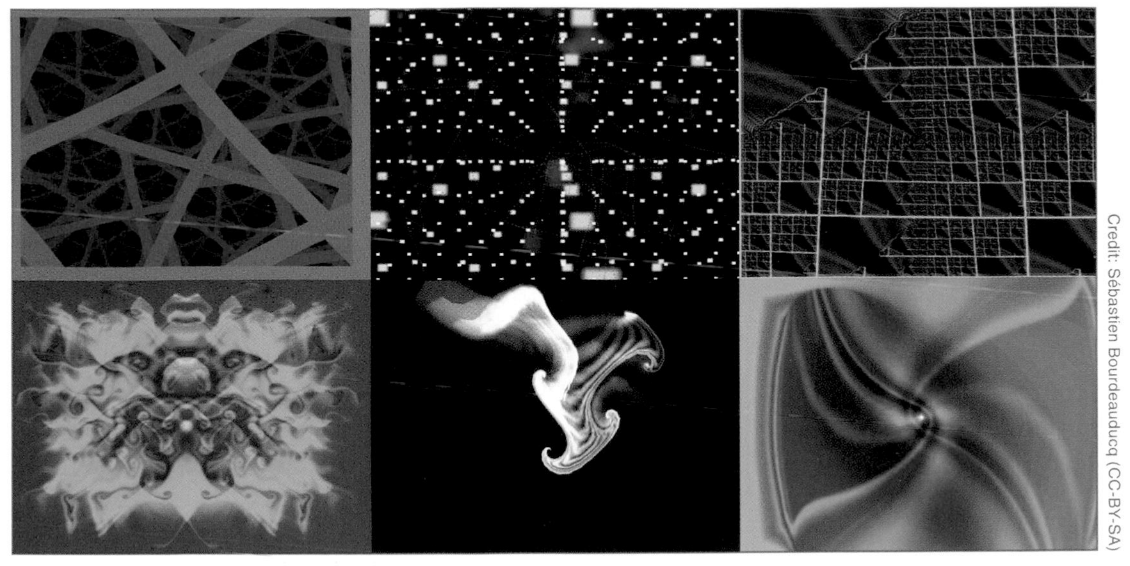

▲ *A sampling of the various video effects created by Milkymist.*

▲ *The Milkymist board serves as the brain of the system, similar to a computer's motherboard.*

"While this is great for building a community of open source developers, hackers are a minority." An additional hurdle is the fact that the Milkymist looks extremely complex, which Bourdeauducq calls a misconception. "Even people with moderate skills but the right mindset manage to carry out useful hacks on Milkymist."

Bourdeauducq ultimately hopes to make Milkymist an indispensable accessory for clubs, bands, and VJs. "We want to sell it in large quantities through retail chains, music shops, etc. In short, we are like a usual electronics device company, but one that applies (and benefits from) the open source principles everywhere possible. Only a self-determined minority of people and other companies would have a real interest in the open source aspects of our work. We believe this is the best way to make large-scale open source hardware happen."

ALT.PROJECT

TOXIC GAS SENSOR

/tmp/lab is located in an industrial area in the suburbs of Paris, near a pharmaceutical factory. One side effect of this neighborhood is a terrible smell. "It smells like rotten eggs at times," hackerspace member Sébastien Bourdeauducq described. "We wanted to monitor the pollution from the factory, more for the fun of it than as a scientific experiment." Bourdeauducq attached Figaro toxic gas sensors to a Linux-based development board and an analog-to-digital converter and wrote a program that published the sensor readings on Twitter. Eventually, as is often the case, the project ran its course and the components were repurposed.

▲ *Faced with a potentially toxic environment outside the hackerspace, /tmp/lab members built sensors to track pollution.*

To learn more, visit:

http://www.tmplab.org/wiki/index.php/Toxic_Gas_Sensor

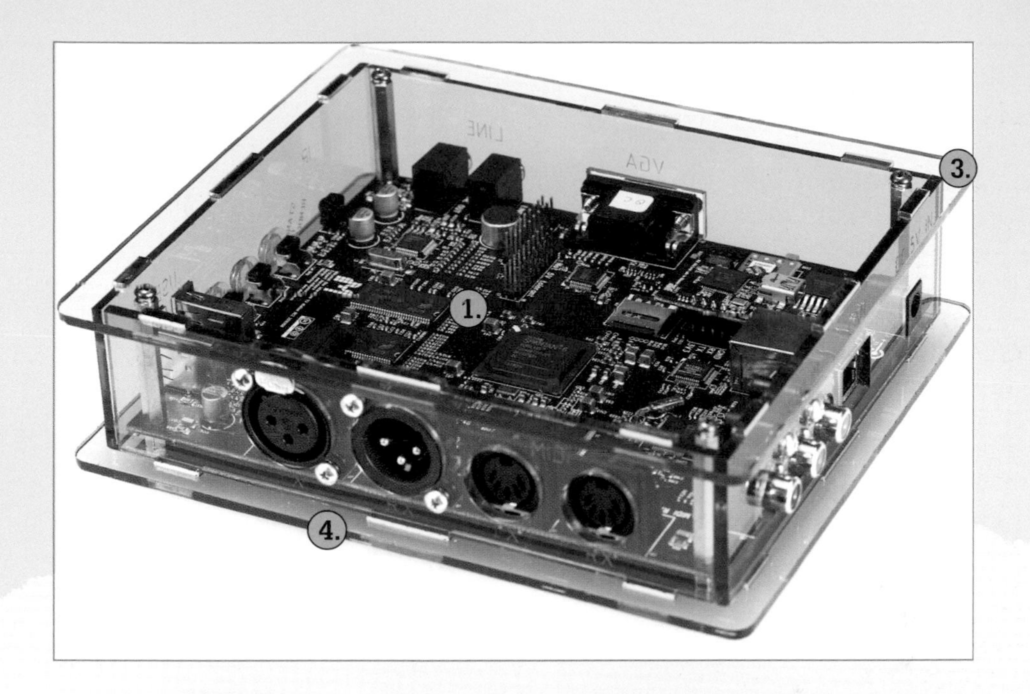

BUILD IT

Outwardly, the Milkymist seems too complicated and challenging to build, but as an open source project, every technical detail and schematic relating to Milkymist and Flickernoise are available for study and modification.

1. **Build the main Milkymist board**—Download the schematics from the project website and manufacture the board. Sending it out to an SMD shop might be the best tactic due to its complexity.
2. **Load Flickernoise software (not depicted here)**—The latest is version 0.4, indicating a beta release that may not be without obvious bugs. The software may be downloaded from milkymist.org.
3. **Assemble the case**—The vector files of the case the Milkymist team created are available from the project sites.
4. **Add the peripherals**—These could include MIDI-compatible synthesizers, cameras, and microphones, among other possibilities.
5. **Throw a party**—VJ the party with your new Milkymist!

FURTHER READING

- Explanation of the manufacturing process—http://en.qi-hardware.com/wiki/Milkymist_One_SMT/DIP_Process_Flow
- File dump—http://www.milkymist.org/mmone/
- Project wiki—http://milkymist.org/wiki/index.php?title=Main_Page

PROJECT 7

WHITE STAR TRANS-ATLANTIC BALLOON

The members of Louisville's LVL1 built an autonomous balloon that can deliver a payload to Europe by navigating the Atlantic jet stream.

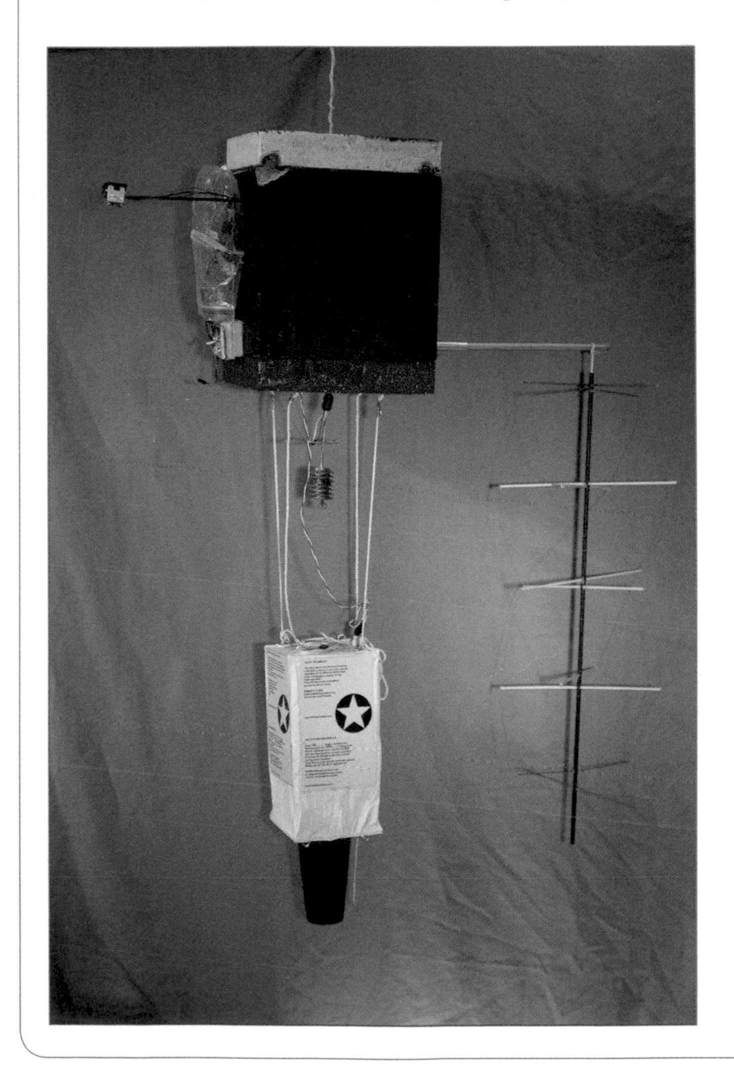

HACKERSPACE PROFILE: LVL1

Location:

LVL1
814 East Broadway
Louisville, KY 40208
www.lvl1.org

Organization type: Kentucky nonprofit

Founding date: November 2009

Number of members: 26

Dues: $50/month

Size of the space: "1,900 square feet officially; 7,000+ square feet unofficially"

Officers/Leaders:

- Christopher "Micro Colonel" Cprek, president
- Brian Wagner, vice president
- Mark Endicott, treasurer
- Tim Hardin, secretary
- Joe Pugh, director at large
- Cindy Harnett, director at large
- Todd Chandler, director at large

Notable equipment: Makerbot, CNC, server rack with full compliment of donated servers, multiple oscilloscopes, power supplies and other electronics test equipment, table saw, MIG welder

SPACE DETAILS

"LVL1 is located in a very large building in downtown Louisville," described president Christopher Cprek. "The previous tenant was a massage parlor. The current owners bought the place about four years ago and completely gutted it. Unfortunately for them (and fortunately for us), the real estate market collapsed, leaving them with a huge, unoccupied building with ornate tin ceiling tiles and skylights. We worked out a deal to rent the back part of the building. The landlord needs a tenant in order to get insurance on the whole place. And because no other tenants are in the space, we effectively get the run of the whole building! Members have built rolling workbenches outfitted with soldering stations and electronics equipment upstairs. Plans are currently underway to add to the existing power tools downstairs to outfit a full fabrication area."

Credit: Kim Gillespie

▲ *Half-finished projects abound in LVL1's main hacking area.*

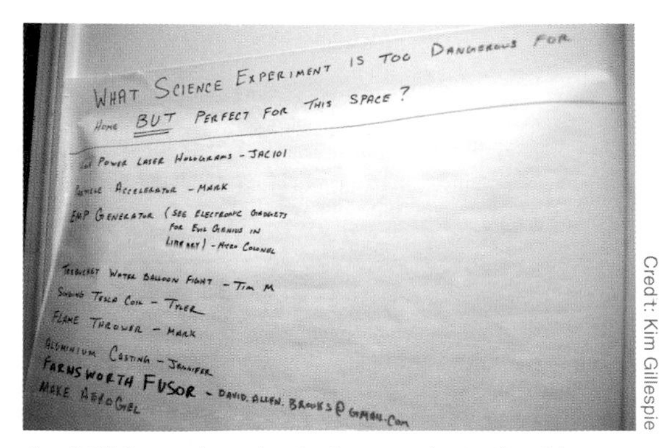

Credit: Kim Gillespie

▲ *LVL1 members brainstorm projects they'd never risk tackling at home, reinforcing hackerspaces' role in unleashing the imaginations of their members.*

THE PROJECT

Participants: Dan Bowen, Tim Miller, Brad Luyster, Jose Cabrera, Tad Heckaman, Tyler Martin, Joe LaGreek, Bill Brown, Aaron VerDow, Bill Piepmeyer, Brian Tanner, Carl Lyster, James Coxon, Nigel Smart, David Stillman, Darrin Embry, Grace Simrall, Gary Flispart, and Tim Hardin

Credit: Kim Gillespie

▲ *A circuit bender modifies a plastic toy to create a customized musical instrument.*

Credit: Nicholas Elrod

▲ *LVL1's members test a weather balloon, taking advantage of the hackerspace's high ceilings.*

Credit: Dan Bowen

▲ *The team built a special cold box that simulates the subzero temperatures the payload will encounter.*

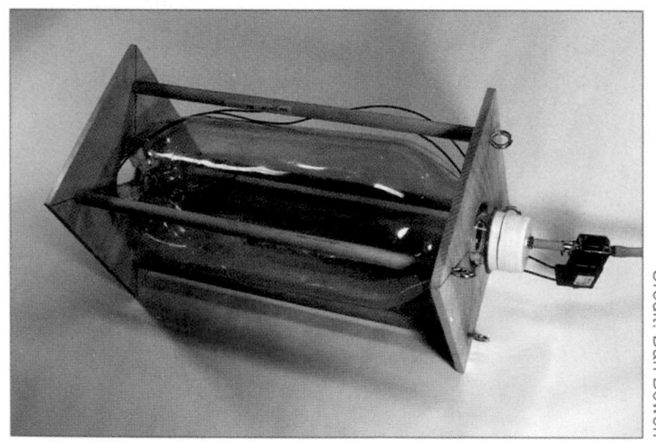

Credit: Dan Bowen

▲ *The balloon's ballast assembly will be filled with alcohol; a solenoid valve releases the liquid on command.*

There have been a batch of hackerspace balloon projects lately; maybe half a dozen groups have launched their own. They mostly follow the same principle: a helium-filled weather balloon, laden with cameras and sensors, rises into the upper atmosphere where the low pressure causes the balloon to burst. The payload parachutes down and is recovered by the ground team. For Louisville's hackerspace LVL1, however, the challenge of getting a balloon into space wasn't nearly as interesting as getting a balloon to Europe.

LVL1's goal was to fly a balloon along the transatlantic jet stream and land it safely somewhere in Europe. "No amateur balloon has ever done this," project lead Dan Bowen declared in a blog post. "The farthest amateur flight so far sank just 200 miles short of Ireland—tantalizingly close, but no cigar." LVL1 call their balloon project White Star.

The core strategy of the White Star team is to equip their balloon with homebrewed sensors and microcontrollers so that the balloon remains within the jet stream. What exactly is the jet stream? Imagine a swirling river of air 2 miles thick; 100 miles wide; and stretching 4,000 miles in length. Its winds gust to 200mph, or as strong as a Category 5 hurricane wind. If that weren't challenging enough, air temperatures can drop as cold as –40° Fahrenheit. All of these conditions make it exceedingly difficult to keep the balloon within the jet stream and functioning for the 36 hours it takes to traverse the ocean.

The White Star project may very well be the most complicated project yet undertaken by a hackerspace. It consists of the following components: the balloon is a 2-mil, Global Western envelope, specified to float at an atmospheric pressure of 250mb, or about

Credit: Dan Bowen

▲ *The team created numerous custom-printed circuit boards for the project, including this PCB, which supplies the payload with electricity prior to launch.*

Credit: Dan Bowen

▲ *Team member Brad Luyster works on hardware in a hacking session at LVL1.*

34,000 feet. An Orbcomm VHF 2-meter satellite data network sends telemetry to mission control while accepting commands in return. A ballast tank holds 6 lbs. of alcohol controlled by a Pneuaire 12-volt solenoid valve. A helium sensor at the top of the envelope determines solar heating, and an ice particle sensor detects cloud density as well as the presence of snow. Sensors measuring the relative humidity and ambient temperature allow mission control to calculate the frost point, the temperature at which frost will form on the balloon.

And what happens when the White Star balloon reaches its destination? "The balloon is equipped with a cut-down device which can be triggered via satellite modem from mission control in Louisville," team member Tim Hardin explained. "The empty balloon will act as a streamer to slow the descent of the payload, and mission control will track its location via GPS. The payload has messages in a dozen or so languages describing what it is and instructions to return it for posterity."

In order for the balloon to reach Europe, however, it must conquer a number of challenges. One of the most difficult hurdles—one that impacts nearly ever system in the payload—is the frigid temperatures of the jet stream, which are often as low as –40°. "The first consideration is the performance of the battery power supplies," Hardin said. "That is, will they supply enough power in their reduced capacity cold state for a long enough time?" Beyond the batteries, will everything else work? "Everything behaves differently when it's that cold," Hardin explained.

Credit: Dan Bowen

▲ *This PCB monitors the balloon's pressure.*

To test the equipment, the White Star team built a custom insulated cold box using dry ice to cool the entire payload to a temperature of –50°C for three days. After the payload emerged, the team made adjustments. They added Teflon insulation to the wires, they painted the payload enclosure black to help absorb solar heat, and they learned to cycle the ballast valve every hour to keep it from sticking.

Eventually it came time to test the technology, and the team assembled a prototype balloon designated SpeedBall-1. "The SpeedBall-1 launch will be the group's first full-scale, long-duration flight to validate many of the technologies and solutions being employed to prepare for the transatlantic flight later this year," team member Seeker wrote in a blog post. "Although Speedball-1 is not expected to cross the Atlantic, it is equipped to rise to about 34,000 feet and travel rapidly out of U.S. airspace to over the

Credit: Dan Bowen

◀ *The payload's battery pack gets its own protective box. A hinged panel pops open to eject drained batteries.*

Credit: Dan Bowen

▲ *The payload, its enclosure open for tinkering.*

Atlantic ocean." After the team collected data on the project's ability to deal with temperature, air pressure, and radio communications, the balloon would crash into the ocean.

Unfortunately, SpeedBall-1's December 18 test launch was cancelled after technical problems cropped up. Both the satellite modem and the telemetry system's GPS receiver failed to connect to their respective satellites. "With no telemetry or backup location tracking, we could not fly," Hardin explained. The team rescheduled for January 8, but that launch was cancelled as well.

Around February 13, the decision was made to transform SpeedBall-1 from a test flight to the actual mission. The reasons for this were twofold: First, the previous cancellations had pushed back the mission to the end of the jet stream season, which is approximately November through April. After April the jet stream would be weakened and traversing the Atlantic would not be possible.

The second reason was that the team learned that another amateur flight was being planned—Cornell University's Blue Horizon project (http://projectbluehorizon.com/). Blue Horizon is a yearly program that is backed by Lockheed Martin and that began in 2008.

"Given that we're a hackerspace with limited time and resources, and started from scratch about a quarter of a year ago, what we've accomplished so far is nothing short of astounding," wrote Seeker in a blog post. Blue Horizon was scheduled to launch on February 21, potentially stealing the glory if they made it to Europe first. The White Star team vowed to beat their rivals to the punch and moved up the launch to February 18.

Credit: Dan Bowen

◀ *In one of the project's numerous test runs, team member Brad Luyster helps test the payload's communication systems.*

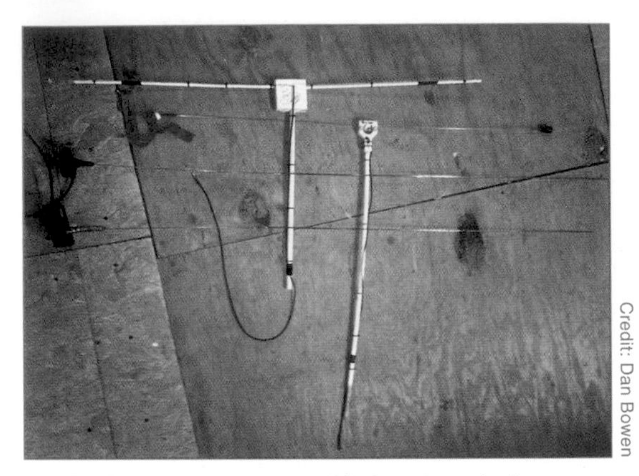

Credit: Dan Bowen

▲ *Radio antennas, assembled and ready for testing.*

Ultimately Blue Horizon scrubbed its flight, leaving White Star home free to make a run for the record. Because the urgency was gone, the team canceled its February attempt and rescheduled it to March 25. Unfortunately, technical problems caused a scrub as well. As these words are written, April has arrived and the team is looking at the possibility that they'll not have all of their technical hurdles solved by the end of the month, necessitating that a launch would have to wait for the following November at the earliest.

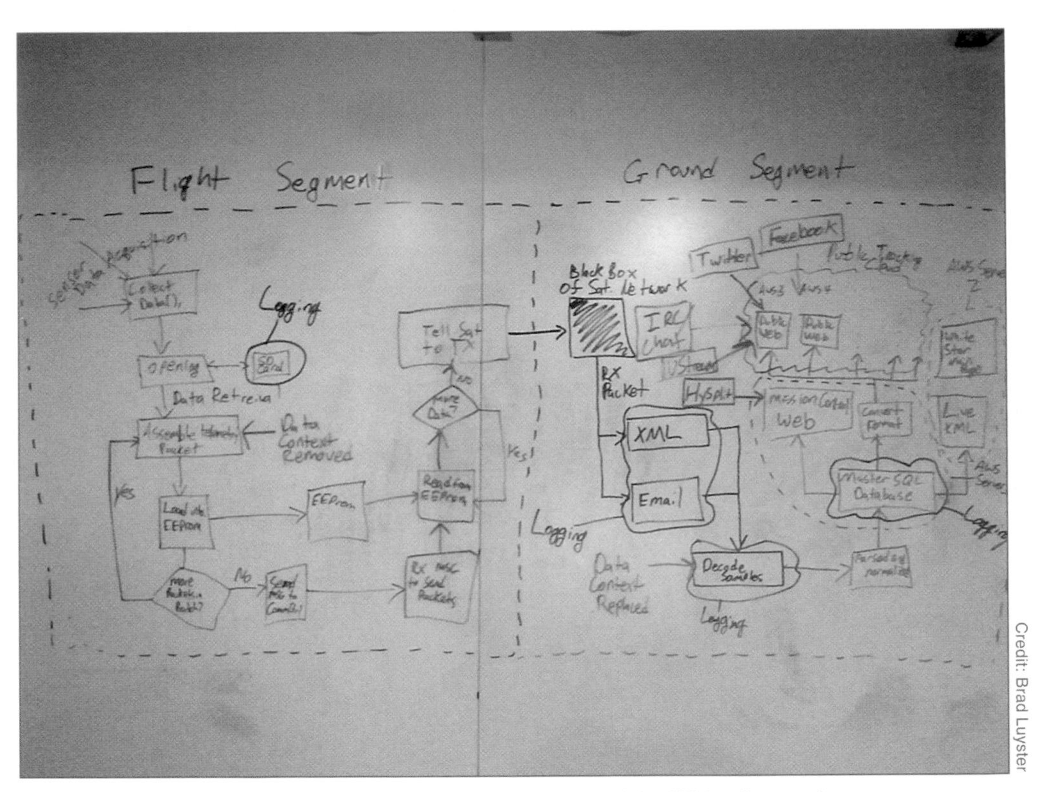

Credit: Brad Luyster

▲ *LVL1's white board shows the incredible complexity of the White Star project.*

Credit: Brad Luyster

▲ *The payload; an XBee module is visible in front. This XBee communicates with a temperature sensor at the top of the balloon.*

White Star's apparent lack of success may surprise a reader, who might have expected a more pat conclusion. The truth is that White Star was a dazzling success without having left the ground. The team developed hardware and software, they created new procedures, and they learned an amazing amount over the six months of the project. Everything they did was documented and shared freely with the world for the advancement of amateur science. In the hackerspace culture, a brilliant failure is second only to a glorious success in importance. In any case, when the next jet stream season begins, the White Star team will doubtlessly be at it again.

HIGH-ALTITUDE BALLOONS

LVL1 is not the first hackerspace to launch its own balloon. In fact, several spaces in the United States and elsewhere have released weather balloons into the stratosphere.

On its most basic level, the act of launching a balloon into the stratosphere is about going into space. For a generation of geeks who were robbed of their dream of being an astronaut, sending a balloon to the edge of the atmosphere carries an undeniable allure. There's also a technical challenge—you're sending a balloon 20 miles straight up, into an area with dangerously low air pressure and temperatures. How do you ensure that you're able to recover your payload?

The classic configuration consists of a helium-filled weather balloon with a parachute and a Styrofoam cooler dangling beneath. The parachute allows for the payload to be salvaged after the balloon bursts in the upper atmosphere, while the cooler holds sensors and cameras. The ground team tracks the payload via GPS, enabling it to be recovered even if it lands many miles away.

Credit: Blake Barrett (Community Commons)

▲ *An image captured by a camera on one of Noisebridge's high-altitude balloons. Numerous hackerspaces have launched balloons over the past few years.*

The most well-known hackerspace balloon program is Spacebridge, a subset of San Francisco organization Noisebridge. Noisebridge, which has its own project featured in this book (Project 24), launched six balloons in 2010 with more on the way, with each mission expanding on the team's capabilities or exploring a new configuration. To learn more about Spacebridge, visit their website at https://www.noisebridge.net/wiki/Spacebridge.

TOOL BIN

DIY PCBS

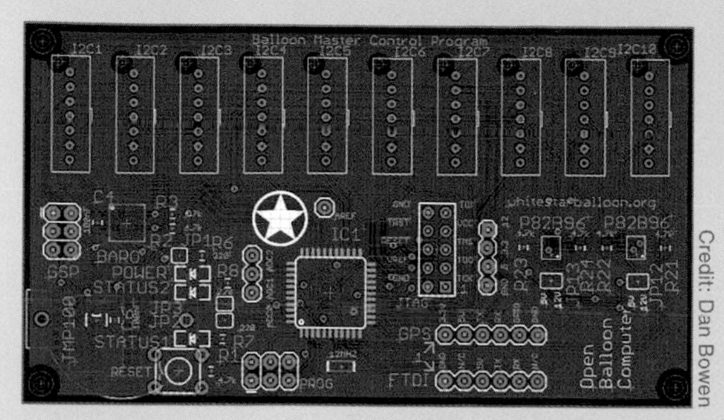

Credit: Dan Bowen

▲ *A PCB design file output by the LVL1 team.*

Credit: Dan Bowen

▲ *Custom boards created by the LVL1 team. After a certain point, it just makes sense to make your own PCB.*

Countless tools have been developed to help electronics hackers build their own circuits, ranging from prototyping boards like the Arduino to hole-drilled perfboard that you have to solder up manually. Eventually, you need a bunch of a certain board, or you want to sell your project in kit form and decide to create your own printed circuit boards (PCBs).

First, you need to lay out the board using design software that creates the circuits. Depending on the software package you own, it can help position the components for maximum efficiency. The most popular hobbyist platform is Eagle, with free and low-cost versions ranging up to a full-featured platform costing hundreds of dollars. However, there are alternatives. "The two popular open source packages are Kicad and gEDA," engineer and kit-maker Adam Wolf described. "Recently, I've been hearing a lot of hubbub about Design Spark PCB."

However, Wolf's favorite is Kicad because it's free, it's open source, and has helpful developers, all while being as easy to learn and use as Eagle and other hobbyist-level tools.

The next step is to output the board as an actual PCB. Some die-hard DIYers prefer to etch their own. The process begins with a copper-clad board. A layer of resist protects the areas that are to remain copper, with the traces drawn by hand or printed on the resist with a photoemulsion. Then the board is dropped into a chemical bath that removes the unprotected copper, leaving the circuit paths behind. "These chemicals are chosen to be able to strip copper off a board in just a few minutes," Wolf cautioned. "It's not something you want to spill, breathe in, or dump down the drain."

Even after the board emerges from the chemical bath, the process isn't complete. "If you are using through-hole components," Wolf explained, "you'll need to then drill all the holes in the PCB using a drill press." The resulting circuits must be checked for continuity using a multimeter. Some DIYers coat the PCB in a tinning solution to help it resist tarnishing and to help solder adhere.

If this sounds like more trouble than it's worth, most electronics hobbyists would agree. An alternative is to use a service that creates the PCBs for you. "There are a variety of companies that will take your artwork files and create a small number of PCBs for a low cost," Wolf said. A customer sends in the design, and the company checks the file for viability and arranges payment. A couple weeks later, the order arrives in the mail. The resulting boards look and work completely professional.

"They're two sided, have soldermask, tinning, and silkscreen labeling," Wolf described. Best of all, the price of these services can be quite low. One of the most popular services, Laen's PCB Order (pcb.laen.org), charges $5 per square inch for three copies of your design.

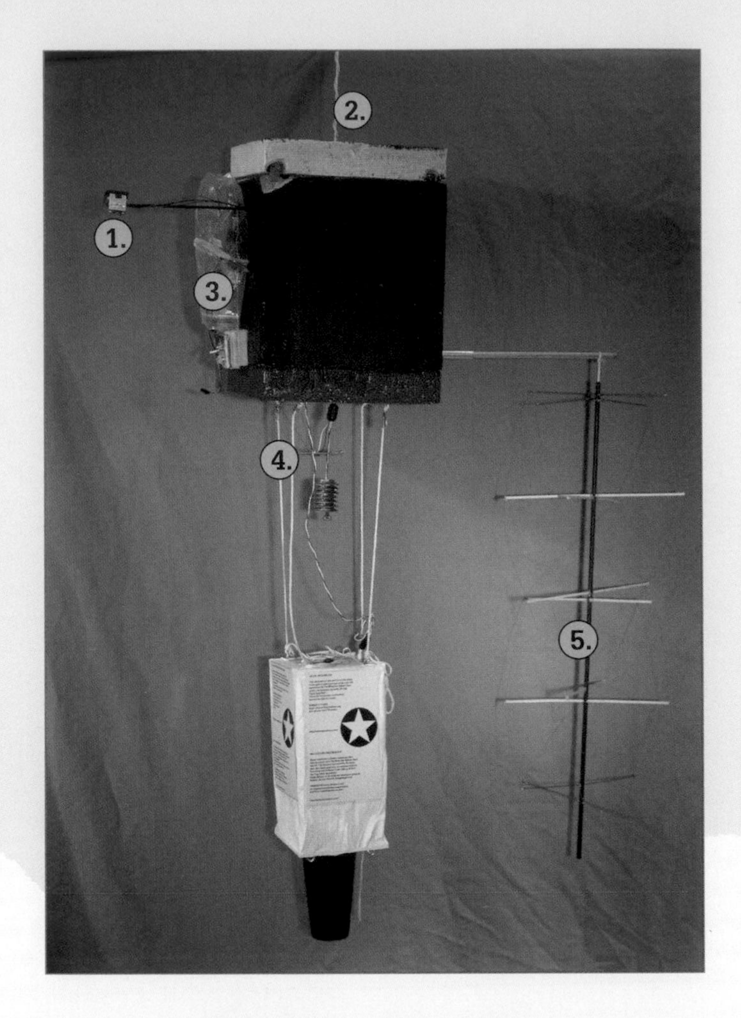

BUILD IT

In addition to the balloon (the White Star team used a 2 mil Global Western envelope) the critical parts of the project are found in the payload:

1. Sparkfun Dust Sensor, used to detect down-drafting ice crystals in jet-stream cirrus clouds.
2. Line to the balloon. A command from Mission Control will activate a cut-down device (not pictured) that severs the line and terminates the experiment.
3. Ham radio shortwave transmitter, donated by a local amateur radio organization.
4. Temperature sensor, covered with a solar radiation shield. The brown PCB above it is a 5-LED strobe that activates in the darkness.
5. Quadrifilar helix VHF ORBCOMM satellite antenna.

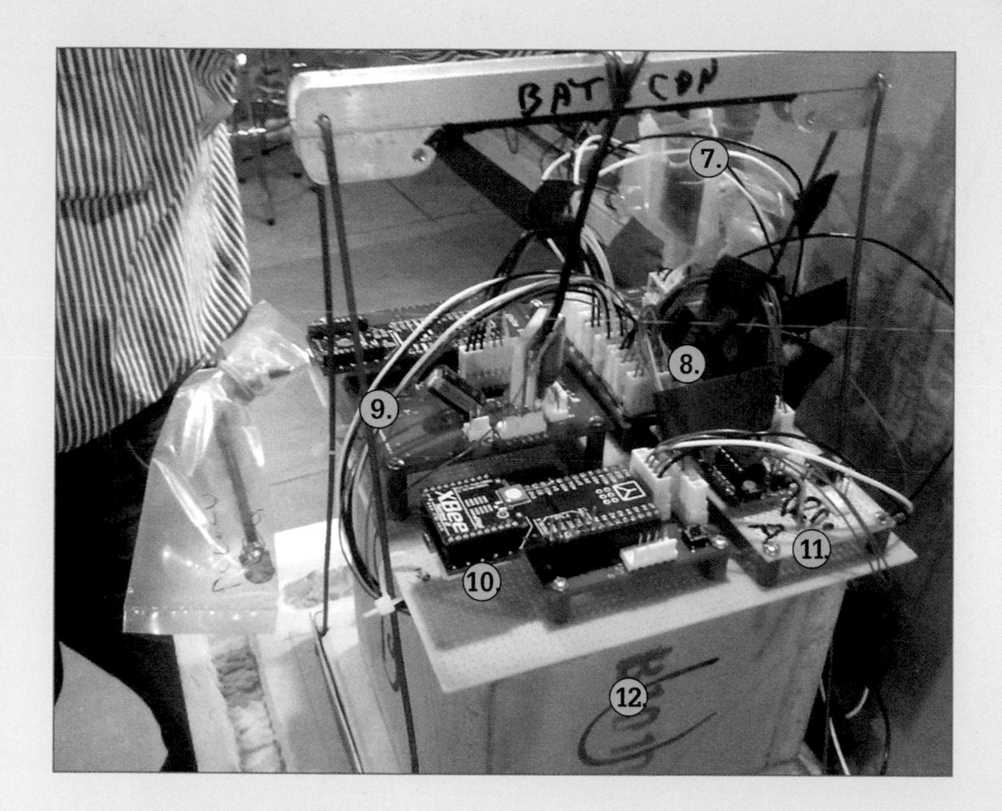

6. The ballast system, consisting of a 3L soda bottle with a solenoid valve.
7. GPS receiver.
8. Flight computer, custom-designed for the project by the White Star team.
9. Communications controller, comprising a satellite modem and an Arduino Fio, which handles message queueing and interfacing with the modem. This assembly receives commands from Mission Control and sends out telemetry.
10. XBee wireless module, which receives data from a temperature sensor at the top of the balloon.
11. Ballast controller, which controls ballast release and reports on how much remains.
12. Battery pack, consisting of 54 AA Energizer Ultimate Lithium batteries arranged into three 18 volt packs.

FURTHER READING

- Main project site: http://whitestarballoon.com/

PROJECT 8

TWITTER-MONITORING CHRISTMAS TREE

Curious to see what topics were trending on microblogging site Twitter, Brooklyn hackerspace Alpha One Labs built a Christmas tree with ornaments that lit up whenever keywords relating to Christmas or their hackerspace popped up.

Credit: Alpay Kasal/Supertou.ch

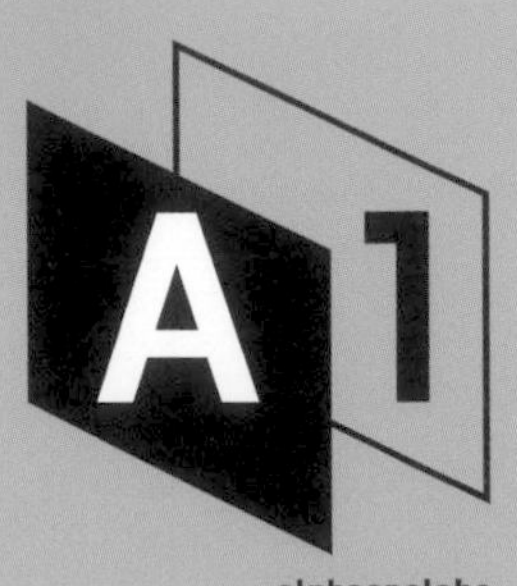

HACKERSPACE PROFILE: ALPHA ONE LABS

Location:

Alpha One Labs
231 Norman Avenue
Brooklyn, NY 11222
http://www.alphaonelabs.com/

Organizational type: New York non-profit, 501(c)3 status pending

Founding date: August 2009

Number of members: 38

Dues: $40/month

Size of the space: 1,600 sq. ft.

Officers/Leaders:

- psytek, member #1
- Mareiska, member #2

Notable equipment: Laser cutter, welder, table saw, circular saw, angle grinder, air compressor, power washer, airless paint sprayer, vending machine, arcade cabinet, soldering irons, oscilloscope, variable power supply, and function generator

Credit: Mitch Altman

▲ *The Alpha One Labs banner adorns one wall of the hackerspace.*

Credit: Jeff Keyzer

▲ *The hackerspace's tool board, known as the Hackerwall, covers an entire corner of the space.*

Credit: Matt Metts

▲ *Alpha One Labs' main hacking area invites members to sit down and start hacking.*

SPACE DETAILS:

Like many hackerspaces, Alpha One Labs is in the process of looking at a new facility. "Our current space is huge; it's actually too big for us," member #1 psytek described. "We originally wanted to redo the floors and ceiling. Once we were halfway through the ceiling project, we realized it would cost too much to finish all of our desired renovations. We've signed on to a new space that has a nice high ceiling, windows, and a great looking bamboo floor. We also have the luxury of having a rooftop deck with panoramic views of the city."

THE PROJECT

Participants: Alpay Kasal, psytek, Robert Diamond, David Prude

In August 2010, a meet-up called a global synchronous hackathon (see sidebar) lured an artist and electronics aficionado named Alpay Kasal to Alpha One Labs. While chatting with the hackerspace's co-founder, psytek, the two conceived the idea of an interactive Christmas tree with ornaments that reacted to specific keywords on Twitter. The plan was to build each ornament with an LED inside, which would light up when the appropriate keyword was detected.

Credit: Alpay Kasal / Supertou.ch

▲ *Individual ornaments light up as Alpha One Labs' tree detects a tweet.*

Credit: Alpay Kasal / Supertou.ch

▲ *The project wiring includes three Arduinos and several transistors. Ultimately, two of the Arduinos weren't needed.*

Credit: Alpay Kasal / Supertou.ch

▲ *A Processing sketch scans Twitter for keywords.*

HANDLES

If you visit a hackerspace, or read about them in a book like this one, sooner or later you'll encounter a member who uses an alias, or *handle*. The tradition of the handle goes back to the screen names favored by the original pioneers of the Internet, who used aliases to guard their privacy and to add a little excitement to their online personas.

The handle tradition has been inherited by hackerspaces, including Alpha One Labs, whose co-founder psytek goes by his alias while in the space.

"I love hanging around other hackers just so we can use handles," he said. "And having a hackerspace where you can use handles on a daily basis is awesome."

Credit: Alpay Kasal / Supertou.ch

▲ *A tangle of discarded LEDs demonstrates the team's change to incandescent bulbs.*

Credit: Alpay Kasal / Supertou.ch

▲ *Decals litter the desktop as the ornaments come together.*

In December, Kasal and psytek reconvened to work on the project. "I started by getting the tree, bringing it over, and routing the wires," psytek said in an interview. Kasal, meanwhile, hunted for clear ornaments to house the LEDs—but couldn't find any.

"I thought this would be the easy part," Kasal wrote in a blog post. "I underestimated the elusive nature of ornaments in the wild. Eventually, I found golf display cases at The Container Store."

They sprayed the cases with Rustoleum artificial frost to make them more "Christmassy" and to help diffuse the light.

They pulled an all-nighter wiring the ornaments, but they quickly realized that the LEDs simply weren't the right solution for lighting the ornaments. They needed the brightness of incandescent bulbs.

"I found some emergency exit bulbs at Home Depot that were rated at 12 volts," psytek recalled. "I picked up 20 of them."

One of the downsides of brighter bulbs was that they needed more volts to drive them. That's where team member Robert Diamond came in.

"Robert helped us pick out transistors and set us up with a working circuit," psytek said. But even that was a hassle.

"We had such a runaround time trying to get the transistors."

Credit: Alpay Kasal / Supertou.ch

▲ *An ornament, powered by twin LEDs, is complete, but eventually the team switched to incandescent bulbs.*

They tried several Radio Shack stores as well as an independent electronics store before they found enough.

The next step was to build the tree's interactive function.

"A netbook running Processing is used for performing searches of predetermined holiday words," Kasal described in a blog post. Processing is a simple programming language based on Java, designed to be accessible to hobbyists and artists. Its programs, called sketches, are run on a computer's desktop and send commands to connected microcontrollers and can also interface with the web. The resulting data generates a queue of ornament flashes.

"Some words came up much more frequently than others," Kasal explained. "The netbook acts like a queue manager, ensuring that the Arduino wouldn't get confused when its own memory got hosed by a high-frequency word, and possibly missing out on a low-frequency word as a result." The Arduino, unlike a full-fledged computer, doesn't have the memory or processing speed to keep up with the flow of data.

In addition to helping with transistors, team member Robert Diamond was pulled in to supply the microcontrollers that managed them.

Credit: Alpay Kasal / Supertou.ch

▲ *The ornaments await activation. The topmost cube, marked 1337, refers to hacker compliment "elite" with its abbreviation "LEET" converted to similar-looking letters.*

Credit: Alpay Kasal / Supertou.ch

▲ *When a Twitter user enters "merry xmas" in a tweet, the appropriate ornament lights up.*

Credit: Alpay Kasal / Supertou.ch

▲ *Another ornament lights up.*

"Robert contributed his uln2803a chips, which helped with the refactoring of brighter incandescent bulbs," psytek said. The uln2803a chips are arrays of transistors with a single chip able to power multiple high-voltage circuits.

Ultimately, the team chose two dozen search terms to feature on the ornaments, including holiday themed keywords such as *santa*, *2010*, and *mistletoe* as well as the hackerspace-related *arduino*, *alphaonelabs*, and *hack*.

The project fermented over the course of about half a week and then pulled together in a span of 24 hours—a few days before Christmas.

"It would have been nice to get it done earlier and not have to have pulled an all-nighter," psytek recalled ruefully.

Despite the last-minute rush, the project was a success, with the tree operational at the space over the holidays, regaling visitors and members with flashing reminders of Twitter keywords.

Even better, the infrastructure has been put in place for future projects of a related nature.

"One project that spawned out of the Christmas tree is our LED sign," psytek said. "We now have an LED scroller that shows the latest tweet mentioning @alphaonelabs."

But they haven't forgotten the Christmas tree and plan to rebuild it every year as a hackerspace tradition.

GLOBAL SYNCHRONOUS HACKATHON

As the hackerspace phenomenon spreads through the world, effort has been made to unite the groups through various events. One of the most popular types of event is the global synchronous hackathon, in which multiple hackerspaces around the world hold all-nighter hacking sessions on the same day, connecting to each other via video streams.

Alpha One Labs' psytek has been at the forefront of the hackathon scene, helping to organize events and even coming up with themes to add a little spice.

"It has been a blast," he said. "Early on we had a bunch of participants and have had members who have joined from coming to the hackathons."

While the hackathons' energy and participation levels have dropped somewhat of late, psytek is still excited.

"We'd like to keep doing the hackathons and want to present a challenge for each one. We are always up to keep the space open for the whole weekend for people to hack away all night until the morning.

Credit: Alpha One Labs

▲ *Video feeds from around the world show the popularity of a Global Synchronous Hackathon.*

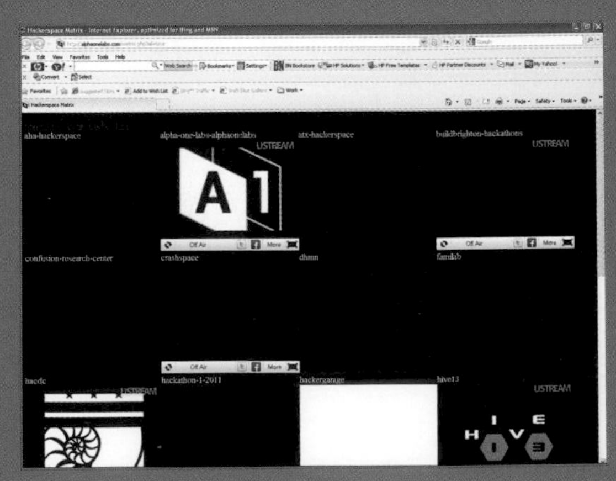

▲ *A video feed for globally synced hackathons. To see the current matrix from the various video feeds, visit http://alphaonelabs.com/matrix.php?all=true.*

TOOL BIN

TRANSISTOR

A transistor is a semiconductor device that has become one of the fundamental underpinnings of modern electronics. Although they come in multiple configurations, a typical unit consists of a wafer of semiconductor with three terminals extending from it. One of the terminals accepts voltage, another outputs this power, and the third controls this process. Because the output voltage can be much powerful than the controlling voltage, transistors are often used to amplify signals. They also work as software-controlled switches enabling a microcontroller to selectively trigger numerous circuits.

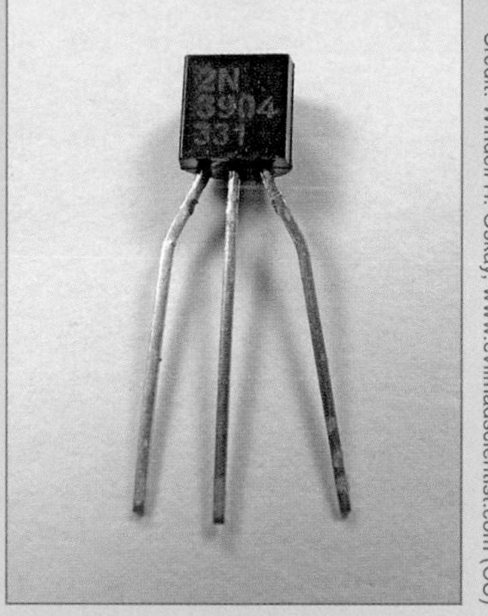

Credit: Windell H. Oskay, www.evilmadscientist.com (CC)

▲ *The transistor is one of the most important components in the tinkerer's toolbox.*

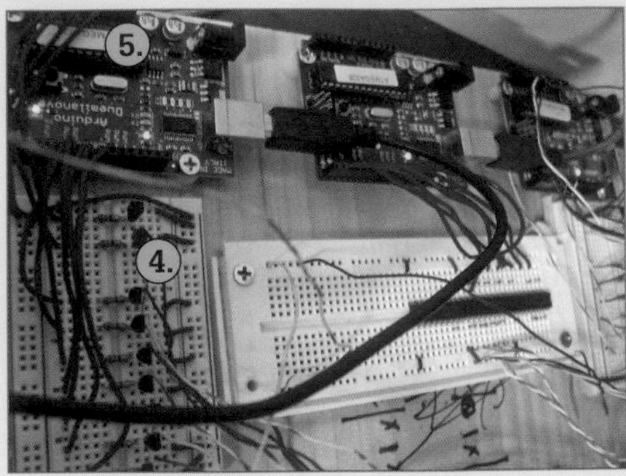

BUILD IT

Building your own Twitter-monitoring Christmas tree is an easy way to add some interactive holiday cheer to your hackerspace. It's also an easy way to monitor what people are tweeting about your organization.

1. Get a tree.
2. Decide how many search terms your tree will search for; you'll need that number of ornaments.
3. Wire an incandescent bulb for each term, and build it a case. Alpha One Labs used golf-ball display cases sprayed with artificial frost to help diffuse the light.
4. Wire the Arduinos and transistors. Each bulb needs its own transistor to drive it and a resistor to protect the transistor.
5. Attach the power supply. A 400-watt or better ATX supply should work fine. Not only does the Arduino need power, but each incandescent bulb does as well.
6. Configure the computer and connect it to the Arduino. You'll need to download the Twitter4J library from http://twitter4j.org/en/index.html. The library is designed for Java, but if you want to adapt it to run on Processing, instructions on the website will tell you how.

PROJECT 9

LIVE WIRE GO GAME

Denizens of Prague hackerspace Brmlab liked the ancient board game Go, but they wanted to add a little more zing in the form of an electric shock that occasionally hits players making a move.

HACKERSPACE PROFILE: BRMLAB (HACKERSPACE PRAGUE)

Location:

Brmlab

(Hackerspace Prague)

Bubenskà 1, 17000 Praha 7

Czech Republic

brmlab.cz

Organization type: Nonprofit with member-elected representatives

Founding date: June 2010

Number of members: 31

Dues: 500CZK, students pay half

Size of the space: 100m2

Officers/Leaders: "I am the chair, representing brmlab in legal matters," Petr Baudiš described. "For dealing with boring day-to-day operational matters, we have the council—me, Dominik Pantůček, Pavol Rusnák, and Pavel Růžička. We also have the revision committee—Michal Tuláček, and Ivor Kollár—which is supposed to watch over us and in practice is helping out with administrative and points out things we might have forgotten about. In practice, we vote on new members after they start visiting the space regularly and we get to know them a bit, try to make members pay dues, and make sure the rent is paid. The really interesting matters like resources used by various projects, access policies, and whatever are decided on meet-ups by general consensus."

NOTABLE EQUIPMENT

Rep-Rap, soldering irons, dremel, etching tools, and lock-picking toolkit. "No laser cutters, CNCs, or fancy stuff like that," Baudiš said.

SPACE DETAILS

"We are a very 'technical' space," Baudiš described. "Most of our projects are related to electronics and computers—not so much going on with regard to 'maker' projects, art, or biohacking, but I hope this will change over time."

Brmlab's visual identity involves computer connectors and ports. "I think the main theme of our space is 'connecting,'" Baudiš said. "We connect people together and connect them with awesome devices and knowledge. I like the hackerspace because it confirms my life's belief—if you are enthusiastic about your idea and have no fear, you can get done even things that seem impossible at the beginning. Probably most of our projects started as 'wow, sounds cool; let's get hacking and see how difficult this is.'"

The hackerspace is situated in an industrial building with a lot of artists as tenants. "We don't have too many art-minded people [in the hackerspace], so painting abstract art on walls is not the kind of thing you will find much. But there are plenty of posters; we are building a semi-hermetic entrance for our fledging biolab; and we have a digital locking system, self-service barcode reader and credit system for beverages and snacks, semi-automated presentation and recording/streaming system, and other things like that.

"One part of our philosophy is to document as much as possible on our wiki, ideally so that I can come up to some gadget and get it right the first time I use it. Another part of our philosophy is 'everything is a project' and while this is not always the case, most of the time we organize things this way, be it cool gadgets or parts of the infrastructure," Baudiš said.

Credit: Felipe Sanches

▲ *The Brmlab crew plays around with a laser projector.*

Credit: Ján Teluch

▲ *The space is working on a hermetically sealed door to secure their biolab.*

Credit: Ján Teluch

▲ *Brmlab hackers have a strong interest in biology.*

Credit: Ján Teluch

▲ *A member works on a RepRap 3D printer.*

Credit: Ján Teluch

▲ *Members and visitors congregate during an open house.*

THE PROJECT

Project participants: chido, Pasky, spark, and rei

▲ *chido's cartoon, the Empty Triangle, inspired the project. http://www.emptytriangle.com/.*

Brmlab member chido was obsessed with the ancient game of Go and drew her own web comic, so it was natural to draw comics about Go. One day her strip, the Empty Triangle (emptytriangle.com), featured a goban, as Go boards are called, that shocked the players. chido decided to actually build the board.

The standard goban is a 19×19 grid with the lines 22mm apart, but oftentimes smaller, 9×9 boards are used to teach beginners or for faster games. chido went with the 9×9 for her electrified board.

Credit: Jaromír Šír

▲ *Make sure you make the right move; Brmlab's Go board shocks you every so often.*

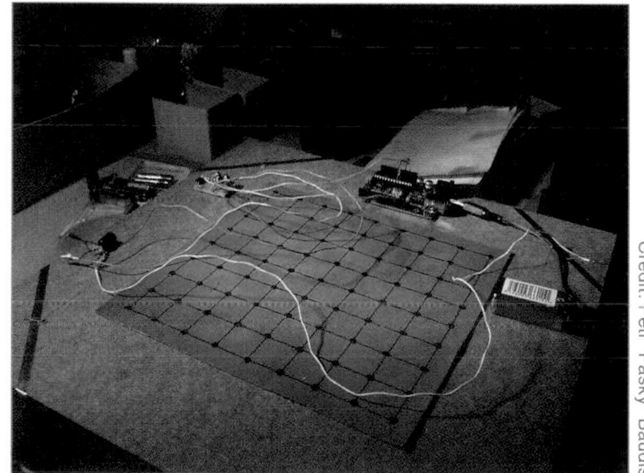

Credit: Petr "Pasky" Baudis

▲ *The game board's top begins to take form.*

The board consists of a wooden box with the Go grid embossed on the top in faux leather, with nails—81 in all—at each intersection. The idea is that a player placing a stone on one of these intersections might close the circuit, thereby receiving a jolt. "When making a move, there always will be a small chance of receiving an electric shock," chido wrote in the project website. "This should give the game a bit of a sadistic twist."

The most important component is the circuit board that dispenses enough electricity to shock players without injuring them. Rather than engineer their own circuit, they found a toy they could repurpose. "We used a ripped-apart shock-your-friend DealExtreme gadget." This is a toy like a fake pack of gum that shocks your friend when you try to grab a piece.

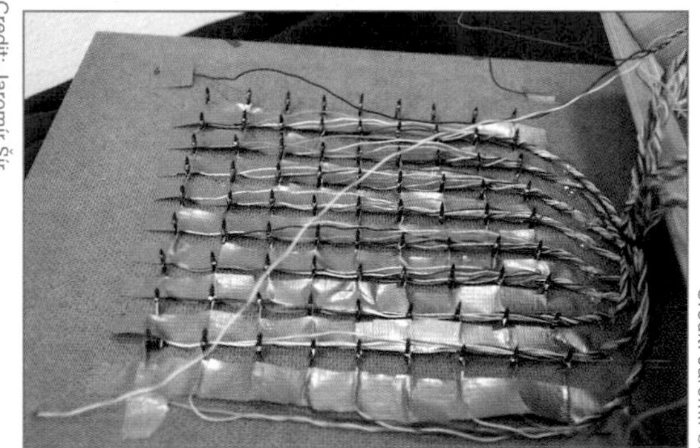

Credit: Jaromír Šír

▲ *The underside of the game's top shows the nails used to create a conductive point at each intersection of the game grid.*

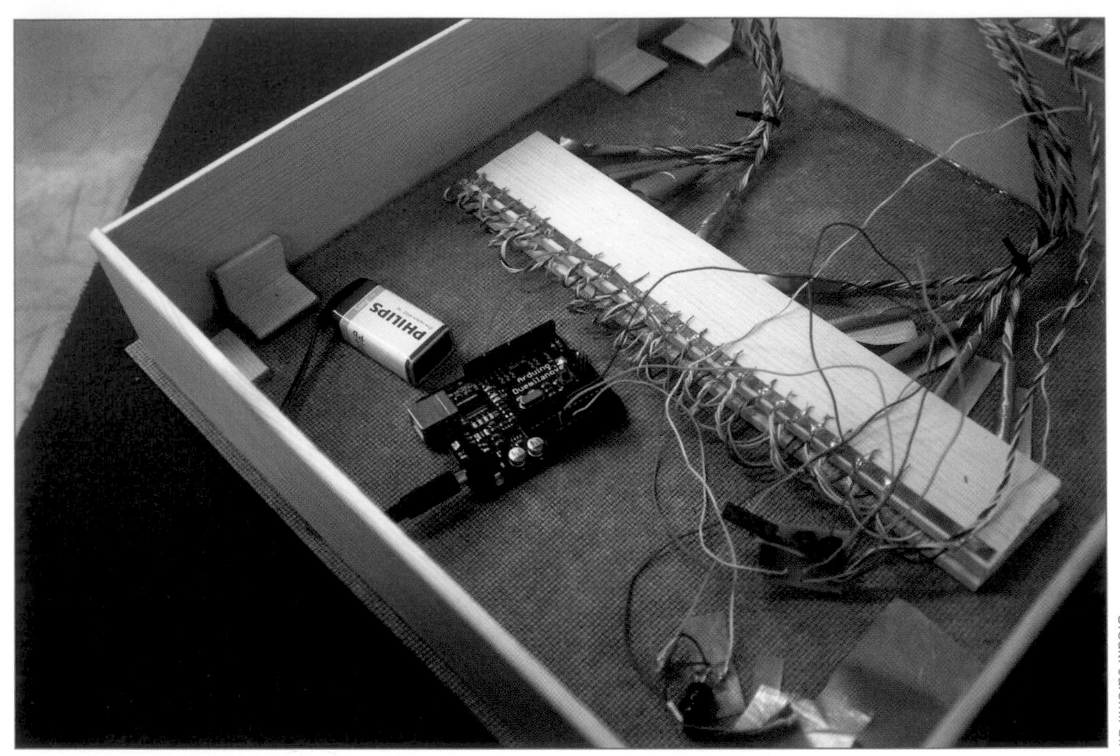

Credit: Jaromír Šír

▲ *The guts of the game include a shocking element from a joke toy and an Arduino to control it.*

Even though they could be confident that the shock wouldn't hurt people too badly, they still needed to figure out how to trigger the electricity only when a piece was being placed. "We decided to design the circuit as having the logic control circuit galvanically separated from the high-voltage control circuit," chido wrote. "This means, there is no actual electrical connection between the two. A very easy way to achieve that is to use a so-called opto-isolator. At the logic control side, there is just a LED, while at the high-voltage side there is a photo-transistor that conducts electricity only when the LED is lighted. So when the high-voltage should be enabled, the logic control just turns on the LED." Everything is controlled by an Arduino.

Of course, for the circuit to be closed, the stones must be conductive as well. After researching conductive paint, chido and Pasky repurposed metallic nail covers, with each stone consisting of two nail covers stuck together with hot glue. "If made properly, the bottoms connecting sufficiently and the stone should be conductive all over its surface," chido wrote.

Finally, the board was ready to test, so they brought it to Deskohraní, a board game tournament (deskohrani.cz) that included a number of Go events. "It was an unexpected success," chido wrote. "Around a dozen people tried it out; some even daring to play more than one game."

chido has a number of modifications she hopes to make, the first being to tone down the shocks. "Even with maximum resistance the shocks are very intense." She wants to add a potentiometer to the board so players can tone down the shocks to an acceptable level. Even more intriguingly, she wants to add the ability to detect stone placement so that games could be recorded or digitized on a computer screen.

Credit: Jaromír Šir

▲ *Ready game pieces and a sly invitation urge newbies to try the board.*

Credit: Jaromír Šir

▲ *As each move is made, the player touches the grounding strip.*

Credit: Jaromír Šir

▲ *Participants at a board game event try the game.*

Credit: Jaromír Šir

▲ *A brave soul tries the game.*

GO

Go is a 2,000-year-old Chinese board game involving black and white game pieces. It is one of the most popular board games on Earth and also one of the most complicated.

A lot of hackers love the game, not only because of the challenge, but also because of the mathematics behind the game. Mathematician John Horton Conway, who created Conway's Game of Life (see page 254), studied Go and formulated two theories on the basis of the game's math.

Some spaces have even formalized their love of the game: Noisebridge in San Francisco holds a regular Go night and the space owns at least two Go boards. (You can see a Go game on the cover of this book, look in the upper-left corner of the table.)

Credit: Paul Sobczak

▲ *This 19×19 Go board shows a nearly completed game. The challenging nature of Go has caught the interest of many hackerspace members.*

To learn more, see http://en.wikipedia.org/wiki/Go_and_mathematics.

ALT.PROJECT

BRMPAW

BrmPaw is Brmlab's variant of the Northpaw (sensebridge.net/projects/northpaw), an anklet that tells the wearer which way is north by vibrating the northernmost of eight vibration motors distributed around the wearer's leg. "The brain adapts to this new input in a few days, integrating it and making use of it subconsciously." The BrmPaw team, led by Jenda, has created their own circuit boards and custom code. They hope to make a cheaper variant of the NorthPaw, and also talked about using it as a haptic clock.

http://brmlab.cz/project/brmpaw

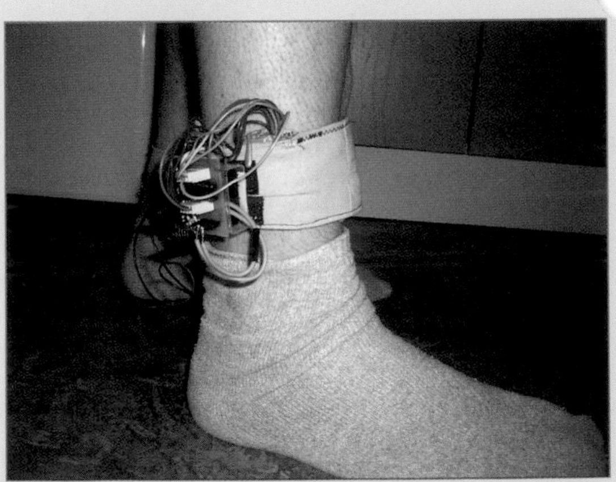

Credit: Jan Hrach

▲ *The BrmPaw anklet will tell you which way is north with a subtle vibration.*

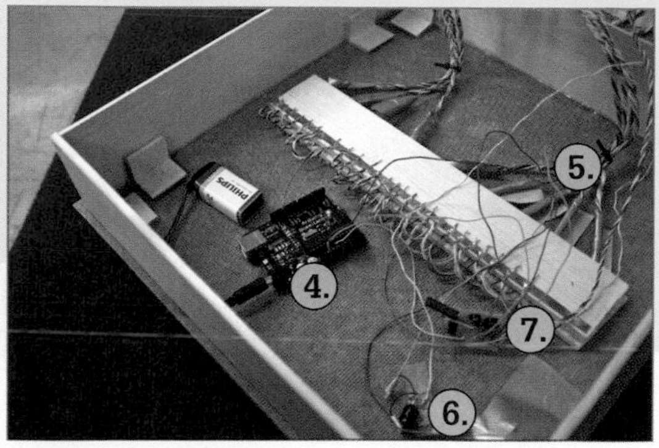

BUILD IT

Building your own electrified Go board can be a cinch if you follow the directions chido wrote on the Brmlab website (brmlab.cz/project/goban):

1. Build case. This project used 35×35cm sheets of particle board with 7.5cm wooden sides.
2. Attach grid. This is a 9×9 faux leather goban nailed to the top board.
3. Add grounding strips. Each corner of the board has a grounding strip that players must touch when they make a move.
4. Connect the Arduino. Instructions and code may be found at the previously mentioned website. In addition, a 9V battery powers the entire game through the Arduino.
5. Wire up the contacts. Each nail has its own wire for a total of 81 wires.
6. Connect the shocker element. This has three leads: one is voltage in, one is voltage out, and the third is a control lead that triggers the voltage. chido and friends used parts from DealExtreme SKU 8572.
7. Add the opto-isolator. This manages the power and turns on the juice only when it detects that the circuit has closed.
8. Make stones. chido and Pasky used copper and aluminum-alloy nail covers and used a hot glue gun to stick two together to make each stone. Be sure to use gloves—the stones get hot!

PROJECT 10

HACK SIGNAL

You know the Bat Signal, the spotlight the cops use to summon Batman when Gotham is threatened? The members of Hack Pittsburgh (abbreviated HackPGH) built their own to summon hackers to the space.

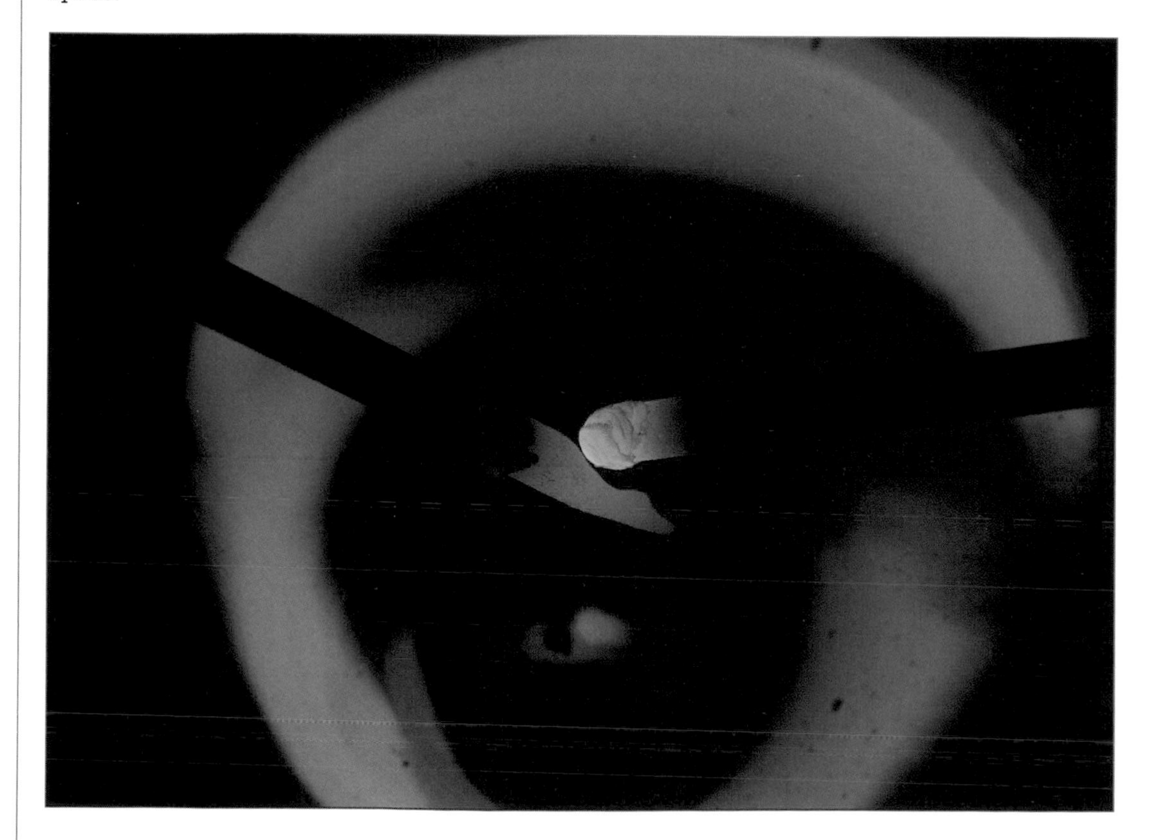

HACKERSPACE PROFILE: HACK PITTSBURGH

Location:

Hack Pittsburgh
1936 5th Avenue
Pittsburgh PA, 15219
www.hackpgh.org

Organizational type: "In transition"

Founding date: March 2009

Number of members: 27

Dues: $30/month.

Size of the space: 1,600 sq. ft.

Officers/Leaders:

- Matt Stultz, president
- Logan Stack, secretary
- Ed Paradis, "council man"

Notable equipment: Soldering irons, electronics test equipment, drill presses, metal and wood lathes, saws (table, band, and miter), air compressor, bench grinder, metal punch, welders

SPACE DETAILS

"We are in a basement garage with a roll-up door," HackPGH president Matt Stultz described. "This makes it really easy for moving things in and out of our space. We came into the space with it being a pretty big mess, but with lots of elbow grease and teamwork we made it our home. The space is under a constant state of development even if we have been there for two years. New shelves or walls are built, things are patched and painted, we keep wanting to improve it."

Credit: Ed Paradis

▲ *HackPGH members work on a weather balloon project (described later in this chapter).*

Credit: Marty McGuire

▲ *Members work on a variety of projects in the space's central hacking area.*

Credit: Matt Metts

▲ *HackPGH members congregate for a talk on unmanned aerial vehicles (UAVs).*

Credit: Matt Metts

▲ *A class on soft circuits shows members and visitors how to add electronics to clothing.*

Credit: Ed Paradis

▲ *Team member Jeff plays with the completed Hack Signal.*

THE PROJECT

Participants: Jeff Kramer, Ed Paradis, Matt Stultz, Eli Richter, and Logan Stack

"I was driving down the street past a new nightclub one night, and they had big spotlights out so people knew where to go to have a good time," recalled HackPGH president Matt Stultz. "I thought, 'Man that's what we need, a giant light to draw people like moths to the shop.'"

After mentioning the project idea to others at the space, Stultz and his collaborators decided to investigate how hard it would be to make their own.

"Jeff [Kramer] loved the idea and started doing some research and found we could build a carbon arc lamp. We did a couple of trials and saw that this was feasible, so we got started."

▲ *An important alert—oftentimes, other members might not be aware of a project's intent.*

The first carbon arc lamp was demonstrated in the early 1800s by British chemist Sir Humphrey Davy using charcoal sticks and a 2,000-cell battery to make an electrical arc over a f4" gap. Before the invention of the incandescent light, most experiments with electrical light involved trying to make the arc lamp practical for widespread use. The harsh white light of this technology—about 200 times as bright as a typical light bulb—made the technology useful only when high-powered lighting was necessary. In addition, the arc's tendency to burn away the carbon meant that the rods had to be continually adjusted or the light would go out.

"Carbon arc lamps work by passing a high-amperage current through two pieces of carbon, a cathode and an anode," Stultz explained. "The high amperage involved causes the carbon to glow red hot and then convert to plasma at the tip of the cathode, before arcing to the anode. The brightest spot of the lamp is not the arc, but actually the plasma-producing cathode tip. This causes all sorts of problems. The plasma is voracious in scavenging free electrons from the air and surrounding molecules, and the compounds produced by the arc are highly corrosive. It also makes a ton of both long and short wave UV—highly dangerous!"

Credit: Ed Paradis

▲ *A carbon rod projects from the side of the garbage can; copper pipes secure the rods. At the base of the can, the steel Ikea bowl is visible.*

Credit: Ed Paradis

▲ *Team members Eli and Matt get the can set up in its cradle.*

Despite these challenges, the HackPGH members decided to tackle the arc lamp.

"We wanted a high-power spotlight, and we had a high-amperage DC welder," Stultz said.

The hackerspace's arc welder would supply the electricity necessary to create the plasma arc.

"Putting the two together required just an initial minimum outlay of a few dollars for arc gouging rods, and we were making light! We decided to continue with it because archaeo-tech is cool, high-powered lights are generally super expensive, and nobody was doing anything like it."

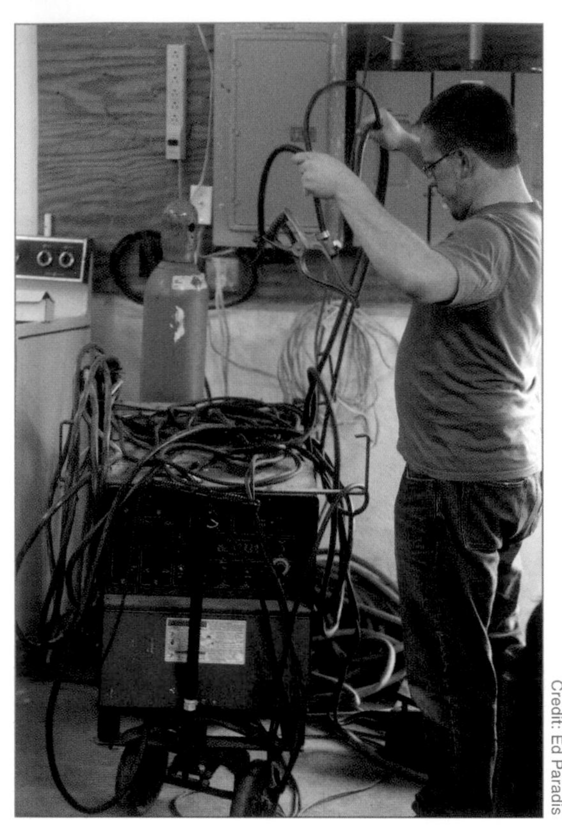

Credit: Ed Paradis

▲ *Jeff wrangles the wires that will deliver electricity to the HackSignal. The blue box is a high-amp power supply used for a welder.*

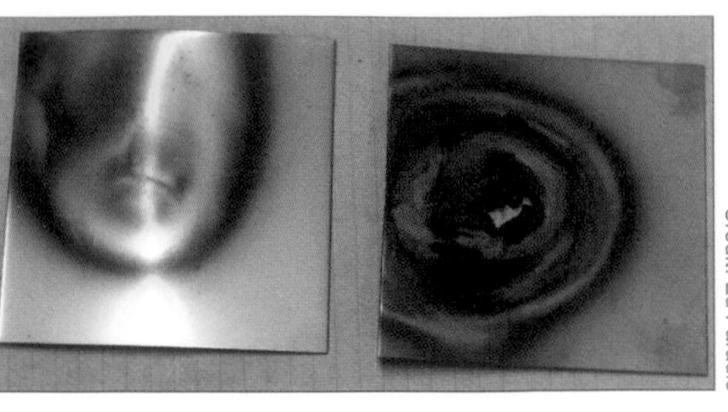

Credit: Ed Paradis

▲ *Steel test plates of different thicknesses; neither proved suitable for use as a reflector.*

Credit: Ed Paradis

▲ *A deceased bumblebee shows the power of the spotlight.*

The team bought some carbon arc gouging rods from their local welding supply store, picking up a few different widths to see which would work. They ultimately went with 1/2" rods. They used a classic zinc-coated steel garbage can as the main enclosure with copper pipes holding the rods.

One challenge the team encountered early on was creating the right sort of reflector to focus the light into a spotlight.

"The first attempt at a reflector was with aluminum flashing," Stultz recalled. The aluminum wasn't able to survive the heat of the plasma and the plasma's tendency to scavenge oxygen from nearby materials. We then moved on to creating the reflector out of thin steel plate."

The team researched how to build a parabola out of separate pieces of metal, cut the steel into strips, and then riveted them together. This burned up quickly.

"Our second reflector was an Ikea salad bowl, and while we need to work on focus, it held up fairly well," Stultz said.

While the space has a functioning spotlight, the dream is to project the HackPGH logo onto a cloud like the Bat Signal of Batman fame. To make this happen, however, they must overcome their focusing problem with a better and more durable reflector.

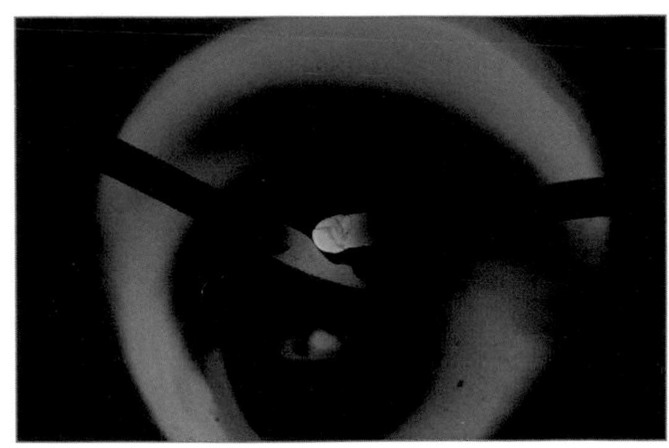

Credit: Ed Paradis

▲ *The glowing carbon rods; the arc is created between the two rods.*

Credit: Ed Paradis

▲ *The light blasts forth, illuminating the night sky—and the wall nearby.*

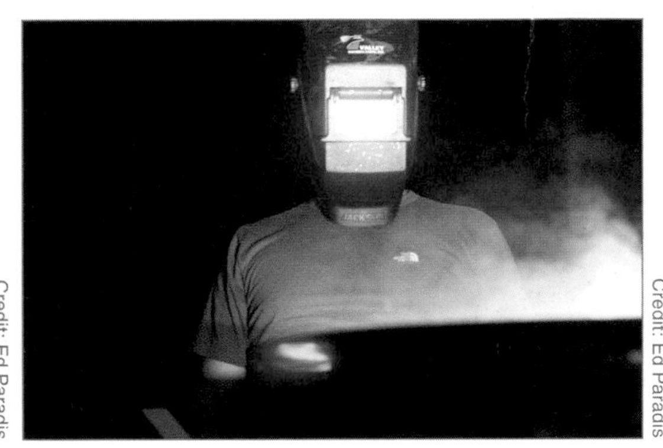

Credit: Ed Paradis

▲ *The spotlight is so bright that anyone close by must wear a welder's mask.*

ALT.PROJECT

AERIAL PING PONG DROP

Using leftover helium from their Near-Space Balloon Project (see page 116), HackPGH released toy balloons into the atmosphere, each packing an origami butterfly and a ping-pong ball. The group used a special machine called an EggBot (http://egg-bot.com/) to print a URL on the balls' curved surfaces, directing the finder to enter her latitude and longitude into a form. While over a hundred balls have been released, only a handful have been found, revealing that the balloons have traveled as far as Erie, PA and Utica, NY.

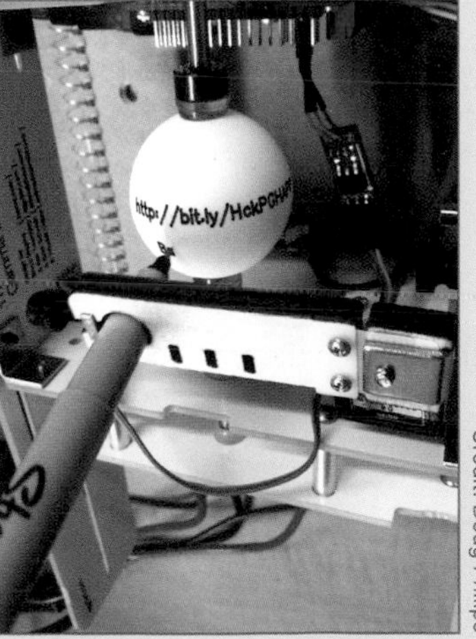

Credit: Doug Philips

▲ *HackPGH's EggBot marks each ping-pong ball with a logo, URL, and unique serial number.*

ALT.PROJECT

NEAR-SPACE BALLOON

In August 2010, HackPGH sent a helium-filled KCI-1000 weather balloon into near space, about 20 miles up. They called the project the LEAD Balloon, with the acronym standing for "lofted electronic aerial device." The balloon carried with it a Styrofoam cooler with a camera and tracking device.

Credit: Mandy Stultz

▲ *The balloon (the white material) awaits inflation. The orange fabric is the parachute, and the box is a Styrofoam cooler holding the payload.*

Credit: Mandy Stultz

▲ *Launch preparations continue as the balloon is fully inflated.*

Credit: Hack Pittsburgh

▲ *One of the photos taken by the Canon PowerShot carried aloft by the balloon.*

HackPGH is not the first space to conduct this experiment and probably won't be the last. It's a challenging but doable project that lets amateurs actually send sensors and recording devices to the edge of space.

The way it works is that the team releases a helium-filled weather balloon, a dangling parachute, and a Styrofoam cooler. The balloon rises swiftly until it reaches the edge of the atmosphere, until the pressure differential between the balloon's gas and outside environment causes the balloon to burst. The payload—protected from cold by the cooler—then parachutes safely down to earth where it is recovered by the team, who follow the signal of a radio tracking device placed in the cooler.

The LEAD Balloon was launched from Perkins Park in Akron, Ohio, and rose to a height of nearly 100,000 feet. They took some photos with the onboard Canon PowerShot set to snap pictures at a set signal; then the balloon burst. The payload was recovered two-and-a-half hours later from a rural area about 25 miles from the launch site.

- **Project page**—http://www.hackpittsburgh.org/wiki/index.php?title=A2_Aug_8th_2010
- **Photos of the launch**—http://www.flickr.com/photos/17267190@N02/sets/72157624683768174/

BUILD IT

If your hackerspace needs a spotlight to show the world that it's open for business, here's how:

1. Drill holes in a metal trashcan. Get a metal trashcan and drill holes for the gouging rods and other attachments.
2. Attach the copper pipes. Attach the copper pipes that hold the gouging rods.
3. Build the reflector. The HackPGH team tried everything from strips of aluminum to an Ikea bowl.

4. Connect the power supply. The team used a high-amp welding power supply that provides enough power to trigger the arc.

Project page:

http://www.hackpittsburgh.org/wiki/index.php?title=HackSignal

PROJECT 11

TARDIS PHOTOBOOTH

Four members of Chicago's hackerspace Pumping Station: One decided to build a replica of the TARDIS, a time machine from the *Doctor Who* television show. As a bonus, they made it into a photo booth. They call it Project Timelord.

Credit: Anne Petersen

HACKERSPACE PROFILE: PUMPING STATION: ONE

Location:

Pumping Station:One (PS:One)

3354 N. Elston

Chicago, IL 60618

www.pumpingstationone.org

Organizational type: Illinois nonprofit

Founding date: September 2008

Number of members: 60

Dues: $70/month for full membership, $40/month "starving hacker"

Size of the space: 2,456 sq. ft.

Officers/Leaders:

- Anne Petersen, president
- Nathan Witt, vice president
- Patrick Callahan, secretary
- Ishmael Rufus, treasurer
- Sam "Rhys" Rees, CTO
- Shawn Blaszak and Jim Burke, directors-at-large

Notable equipment:

MakerBot, welder, oscilloscopes, and a 4-color screen printer

Credit: Anne Petersen

▲ *This is the main hacking area of PS:ONE. The centerpiece is a loft complete with a couch, TV, and bookshelf.*

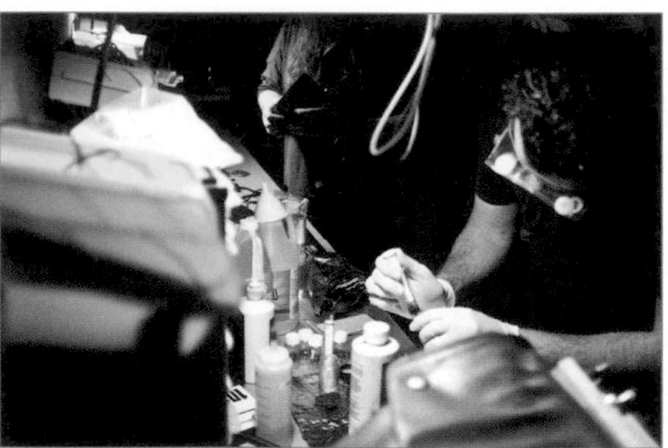

Credit: Anne Petersen

▲ *Two hackers experiment in PS:One's chemistry lab.*

Credit: Anne Petersen

▲ *PS:ONE member rogueclown splits time between a Netbook and an ancient Commodore 64.*

SPACE DETAILS

"Pumping Station: One exists to bring the relationship of art and technology closer together," Nathan Witt explained. "In our two years of existence, our members have joined together on many projects like the Power Racing Series at Maker Faire Detroit, and The VIMBY Challenge & Synchronous hackathons," Witt said. "The culture of PS1 is a 'do-ocracy.' Members are encouraged to take initiative without anyone else's approval to help the space grow and improve." (See Project #23 to learn more about the VIMBY challenge.)

Credit: Anne Petersen

▲ *An (edited) exhortation urges members to tackle challenges without fear of failure.*

Credit: Anne Petersen

▲ *The TARDIS photo booth has become a symbol of Pumping Station: One, as well as a fixture in its main gathering area.*

THE PROJECT

Participants: Jordan Bunker, Josh Kreuger, Nathan Witt, and Anne Petersen

Josh Krueger, a DJ, wanted a photo booth for parties he worked, so he and fellow PS:1 member Jordan Bunker began mocking up the booth's footprint on the floor at the hackerspace. "The two of them decided a cramped little photo booth wasn't going to be enough," described Nathan Witt. "They wanted to party in the photo booth, a party within a party." Bunker looked upon the square masked out on the floor and said, "That's about the size of a TARDIS."

The TARDIS is the personal residence and vehicle of the title character of television's *Doctor Who*—it looks like a British police call box on the outside, but inside it's a huge structure. In addition to being bigger on the inside than the outside, the TARDIS is a space ship and time machine.

Credit: Anne Petersen

▲ *PS:1 member Ryan Lanham assembling corner supports.*

"We live and die by the do-ocracy," Witt said. They split up the roles: Bunker researched the movie prop in order to design an authentic replica. Krueger bankrolled the whole project and spun tunes. Witt began building the TARDIS, while Petersen took photos and helped lay down some of the five coats of paint.

The box was built mainly from 1×4s and sheets of 1/8" plywood, using PS:1's Shopsmith Mark V (see sidebar), a Ryobi TS1342L miter saw, and a Ryobi table saw. It took about three weeks to complete.

Because of the sheer size of the booth, it was designed to be modular. "The final structure starts with a slotted base," explained Witt. "Then you put in posts and walls and finish off with three layers of rooftop."

Credit: Anne Petersen

▲ *Hackerspace members begin painting the assembled wall panels.*

Credit: Anne Petersen

▲ *Power cables protrude from the unfinished walls of Project Timelord.*

Credit: Anne Petersen

▲ *The touchscreen's enclosure takes shape.*

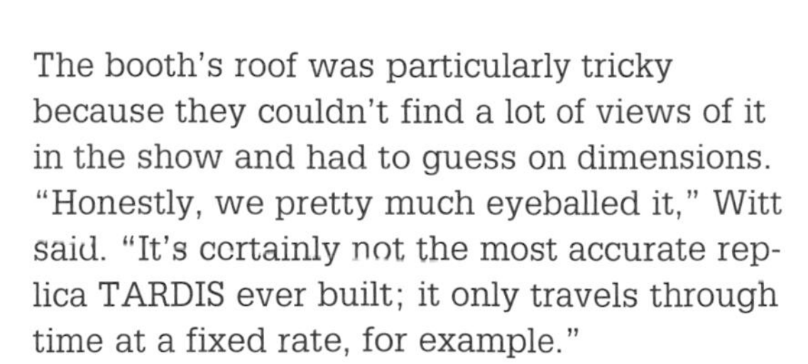

The booth's roof was particularly tricky because they couldn't find a lot of views of it in the show and had to guess on dimensions. "Honestly, we pretty much eyeballed it," Witt said. "It's certainly not the most accurate replica TARDIS ever built; it only travels through time at a fixed rate, for example."

Pictures are snapped by a digital camera. Its shutter is controlled by a Processing application running on a tablet PC. However, Witt hopes to create a more sophisticated system. "I'm in the process of writing an Adobe AIR app that will make use of a webcam," he said. In addition to streamlining the interface, the camera will publish the resulting photos on Twitter.

The booth proved to be a hit and has come to serve as a symbol of the space, getting mentioned in articles and news reports about the space. "It lends us a hearty helping of nerd cred with new people that come to visit us," Witt said. One disappointment? As it turns out, even as modular as the TARDIS is, taking it places isn't particularly practical. "Most of us are city dwellers, and for its size, you do still need an Econoline or something similar to move it. So it makes its home at the space most of the time."

THE ETYMOLOGY OF PUMPING STATION: ONE

Many hackerspaces have relatively unimaginative names, the result of committees that choose names agreeable to the majority while offending the fewest number of people. Usually they end up choosing something like Hack: (City Name) or (City Name) Hackerspace. So, how did this organization end up with a name that is both cool and cryptic?

"The guy who planted the seed of [a hackerspace] in Chicago already had the name in mind before we began meeting," Witt said. "It's never been officially explained." However, it could be a reference to the Great Chicago Fire of 1871. "The old water tower and Chicago Avenue pumping station are a few of the only surviving buildings."

▲ *The finished booth has been a hit at PS:1 events as well as a fixture of the hackerspace's main workroom.*

▲ *With just a couple spots of paint to touch up, Project Timelord nears completion.*

Credit: Anne Petersen

▲ *Ironically, Project Timelord's role as a photo booth has been pushed aside by its popularity. Here, a PS:1 visitor poses for a photo by the space's new landmark.*

TOOL BIN

SHOPSMITH MARK V

Pumping Station: One's Shopsmith Mark V is a 5-in-1 woodworking machine that combines the functions of multiple power tools into a single unit with a 1-1/8 hp motor that powers all the components. It works as a lathe, table saw, drill press, disc sander, and band saw, taking less than a minute to change from one tool to another. Indeed, the Shopsmith has only one motor, so when you start the band saw, the disc sander begins spinning as well. The Shopsmith's ability to pack a lot of features into a relatively small space makes it ideal for hackerspaces, many of which do not have a lot of room to devote to a woodshop.

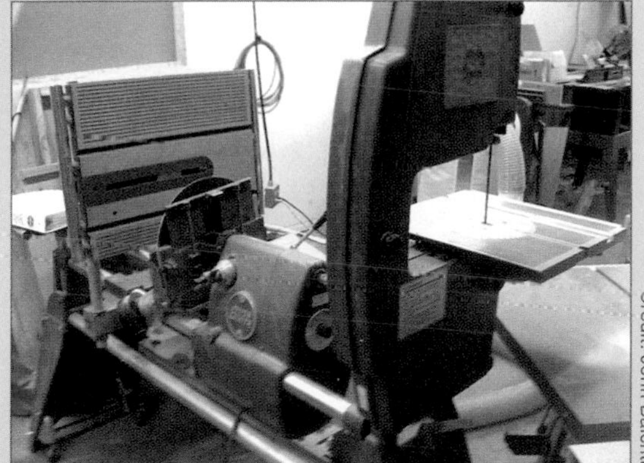

Credit: John Baichtal

▲ *The Hack Factory's Mark V is the centerpiece of the wood shop.*

BUILD IT

If you have a pressing need for a photo booth built in to a replica science fiction time machine, you've come to the right place.

1. Find blueprints or create your own design. Both are easier than you might think. As a staple of popular culture, dozens of *Doctor Who* fans have built their own TARDIS replicas and have published their techniques online. The PS:1 hackers created their own design, eyeballing dimensions while watching the show.
2. Cut and sand the boards used for the sides, floor, and roof. Note that ultimately, you'll want the TARDIS to be sectional so that it can be broken down for transport. Otherwise, it may be too heavy and unwieldy. Even collapsed, Project Timelord is so heavy that it rarely leaves PS:1's common area.
3. Build the sides, door, and back. Then assemble the TARDIS to make sure everything fits before you add the roof.
4. Build the roof. This is typically done in sections due to its complicated architecture. The roof also houses all the electrical and electronic systems except for the camera and touch screen.
5. Wire the lights. In addition to the flashing light on the roof, which the group created with a dimmer switch, Arduino, and servo, the "Police Box" signs also light up with the flick of a switch. Power is supplied by a 120v line that runs through the base and up the wall to the roof.

6. Create the enclosure for the photo-taking mechanism. The interface should be as simple as possible so visitors won't get confused, and to prevent picture-takers from accidentally damaging the equipment. PS:One covered the digital camera with a wooden box and provided a touch screen for guests to control the camera. However, there are many ways to snap pictures. Don't forget to cut a hole in the back of the TARDIS to admit power and data cables.

7. Paint the TARDIS. The official blue color is so dark you'll need at least five coats to ensure that no distracting streaks show through.

8. Done! Sit back and enjoy the reactions of *Doctor Who* fans as they spot the TARDIS and crowd inside to snap their pictures.

For additional photos of Project Timelord, see Anne Petersen's Flickr page at http://www.flickr.com/photos/opacity/sets/72157623022637590/with/4381534818/.

PROJECT 12

INTERACTIVE SPACE INVADERS MURAL

With an eye toward decorating the hackerspace, Metalab members designed an illuminated display of Space Invaders starships that could be programmed to display a variety of colors and patterns.

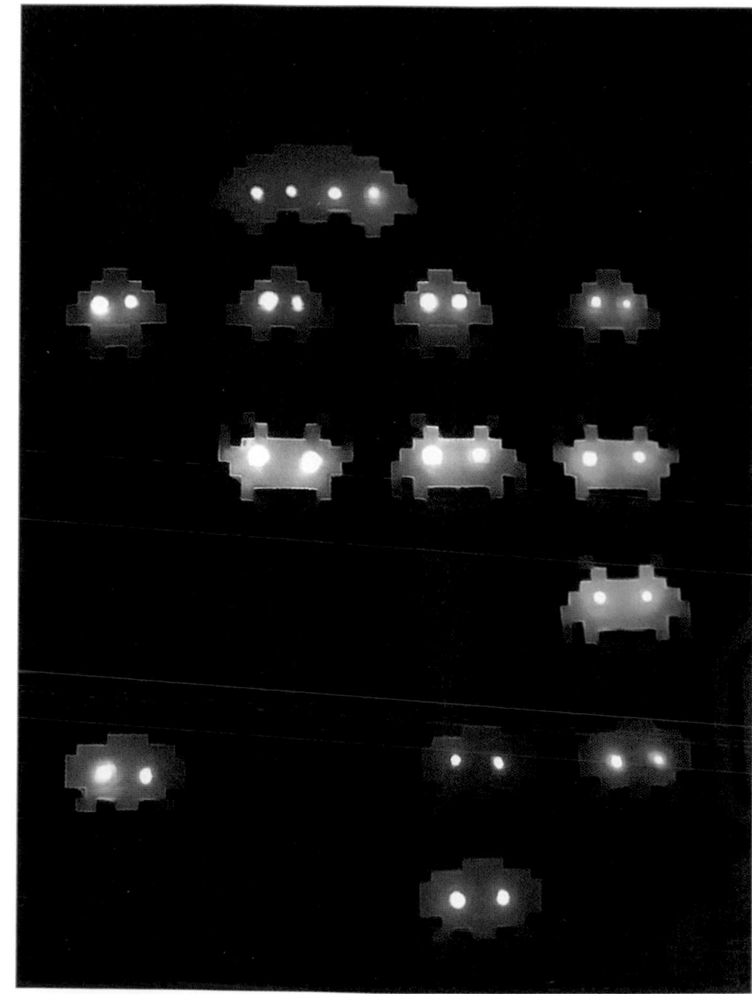

Credit: Phillip Tiefenbacher

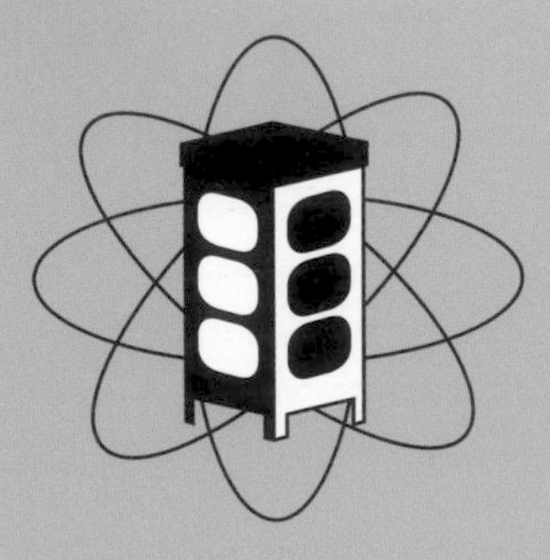

HACKERSPACE PROFILE: METALAB

Location:

Metalab

Rathausstraße 6

1010 Wien, Austria

metalab.at

Organizational type: Nonprofit

Founding date: February 2006

Number of members: 120

Dues: 20€/month

Size of the space: 200 sq. meters

Officers/Leaders:

There are three *vorstandsposten* (board members) mandated by Austrian law:

- Markus "Fin" Hametner: obmann (chairman)
- Sabina "Vandebina": Simonič: schriftführer (secretary)
- Veronika "Bruin" Liewald-Fuchs: kassier (teasurer)

Notable equipment:

Laser cutter, CNC mill, electronics lab, darkroom, chemistry lab, and wood shop. "Also there is a MAME machine with some thousand arcade games on it and a soccer table," described member Florian Bittner.

SPACE DETAILS

"It's big, colorful, blinky, sometimes loud, mostly calm," said Bittner. "Me personally, I enjoy the atmosphere of people working on very unusual things. Also, you can have interesting out-of-the-box conversations in the lounge."

The members have heavily customized what was originally a run-down space, remodeling and painting the walls and upgrading the electricity, lights, and air conditioning.

Credit: Christopher Clay, Community Commons

▲ *The main entrance of Metalab welcomes visitors. Like many displays at the hackerspace, the sign's LED illumination blinks and changes color.*

Credit: Christian Benke

▲ *Metalab members congregate for an online gathering called a global synchronous hackathon.*

Credit: Philipp Tiefenbacher

▲ *A Metalab member tinkers with a robotic hexapod.*

THE PROJECT

Participants: Florian "overflo" Bittner, Philipp "wizard23" Tiefenbacher, and Marius Kintel

Like anyone else, hackers love to decorate their workspaces. But far cry from cube-farm decor like kid pictures or puppy calendars, hackerspaces push the limits with complicated and outrageous art. A decoration that requires advanced technical know-how is a plus because it shows off the skills of the hackerspace.

Credit: Christian Benke

▲ *Metalab members enjoy a lasagna dinner in honor of author Robert Anton Wilson, who urged followers to "keep the lasagna flying." 2011 marks the fifth anniversary of this gathering, held on Wilson's birthday.*

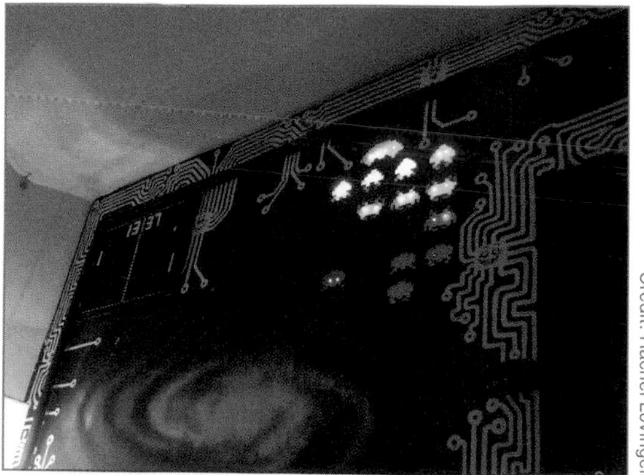

Credit: Rachel Lovinger

▲ *The wall of Metalab's lounge is decorated with a galaxy, circuit board leads, a simulated Pong board, and a set of glowing Space Invaders ships.*

Members of Austrian hackerspace Metalab worked hard to decorate their space, painting the walls and embellishing their main hacking room with stenciled ships from the classic Space Invaders arcade game. One member, metwaly, painted a galaxy superimposed with printed circuit board leads on the wall of their lounge.

Credit: Bre Pettis

▲ *An acrylic Space Invaders ship takes form on the bed of Metalab's CNC mill.*

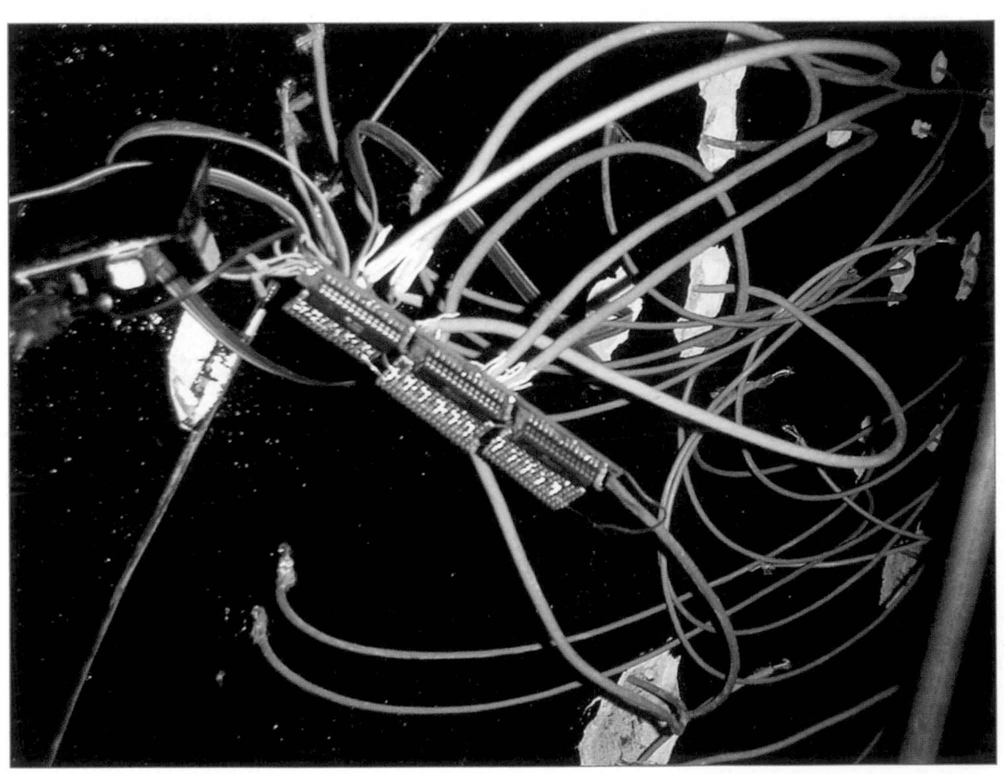

Credit: Philipp Tiefenbacher

▲ *The mural's TLC5940 chips control the LEDs.*

However, Metalab members have gained a reputation for creating what they call "blinkenthings"—decorative patterns of LED lights. Their first such project, called the blinkenceiling, consisted of high-intensity LEDs installed in a hallway, sandwiched between mirror-finished foil and a sheet of white cloth. "It changes color, blinks, has some nice effects, and provides a warm feeling to the room," co-creator Florian Bittner described.

When someone suggested metwaly's galaxy painting needed some Space Invaders, Bittner and collaborators Philipp Tiefenbacher and Marius Kintel created an electronic mural consisting of a series of plastic spaceship shapes illuminated by LEDs.

Bitter described how they divided up the roles: "Philipp is much more into electronics than me, so I did the job of soldering and he did the job of electronic layouts." They split up the coding work while Marius Kintel helped with firmware.

The first step was to mill the spaceships. Bittner found classic Space Invaders shapes on the Internet and created vector paths that the hackerspace's Heiz CNC Step mill cut out from 8mm acrylic.

Credit: Philipp Tiefenbacher

▲ *Team member Florian Bittner solders circuit boards.*

Credit: Philipp Tiefenbacher

▲ *Bittner plugs wires into circuit boards. Soldering the leads in place proved to be a challenge for the team due to tight confines.*

Next, they wired the LEDs, using tEhernet cabling as a way of reducing the tangle of wires. Controlling the LEDs are six TLC5940 LED drivers. These chips expand the Arduino's ability to control LEDs; the TLC5940s collectively manage 28 RGB LEDs—two for each spaceship plus four for the mothership.

The decision to use an Arduino to control the LEDs was not the team's first choice. Originally, they used a PIC, a powerful micro-controller contained in a single chip—but they quickly switched for technical as well as philosophical reasons.

"Arduino is open source software and hardware, as well as multiplatform," explained team member Marius Kintel. "With PCs, you quickly end up using proprietary, Windows-only software, so we left that platform behind for this and virtually all other projects as well."

At the time, TLC5940s were not compatible with Arduinos, so Kintel ended up writing the libraries to make the two platforms work together.

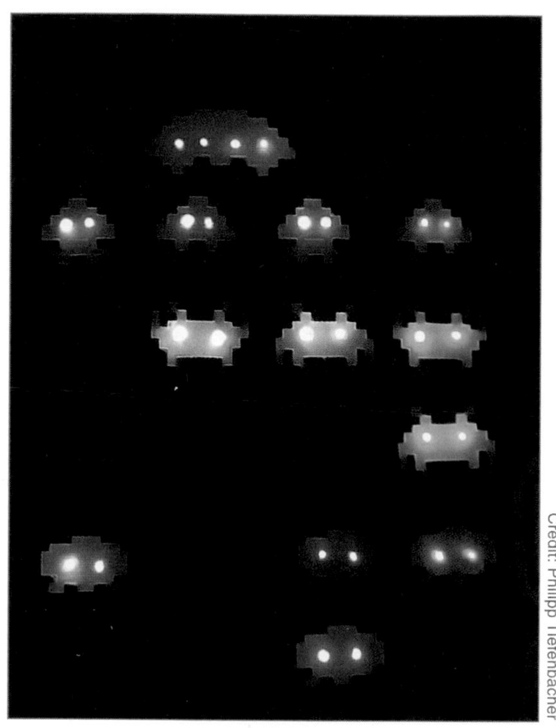

▲ *The Space Invaders ships lit up. With multiple LEDs per ship, interesting gradations were achieved.*

While the finished piece looks very slick, the other side of the wall is a different story. Wires from the LEDs dangle down, held in place with duct tape, while the Arduino and circuit boards are secured to a horizontal board.

"The whole installation is very messy from behind," Bittner admitted.

The completed project has joined other "blinkenthings" to become part of the Metalab, and visitors to the hackerspace routinely snap pictures of the multicolored, blinking alien ships.

TOOL BIN

LED DRIVERS

User-friendly microcontrollers like the Arduino offer a quick concept-to-prototype path, but there are limitations. One of the hurdles Arduino fans continually run into is that only so many components can be plugged into one of these boards.

For instance, if you want to power and control 30 individual LEDs, this is simply impossible with an Arduino. The solution: An add-on board that multiplexes numerous LEDs and allows a microcontroller to individually address them. One of these is the LoL (Lots of LEDs) shield, which provides a convenient matrix of 9×14 LEDs, each of which can be addressed individually, turning the shield into a miniature display. The Metalab team used TLC5940s, which lack the slick interface of an Arduino shield but offer a vast amount of control over the individual LEDs.

▲ *The LoL shield allows the individual control of up to 126 LEDs.*

CLUB-MATE

Hackers have always been known for staying up late or even foregoing sleep in pursuit of a project. Over the years, any number of caffeinated beverages has made this possible, with Mountain Dew, Jolt, and other favorites keeping sleepy eyes open. In Germany, an unusual beverage called Club-Mate came out of the club scene and into the hackerspace.

Club-Mate is a lightly carbonated tea brewed from yerba-maté leaves. It has a very distinctive tea taste that many people dislike at first, so much so that the soda's semi-official slogan is *Man gewiöhnt sich daran,* which translates as, "You'll get used to it." Despite the unusual flavor, many people instantly fall in love with the stuff.

"I was at the 23c3 hacker conference in Berlin on my first trip to Germany," hackerspace founder Nick Farr recalled. "I was fiending for coffee to fight off the jet lag when someone handed me a Club-Mate. I actually liked the flavor from the start, but I really fell head over heels for it when I realized I was fully functional, alert, and not wired-awake."

That is the allure of Club-Mate: it keeps you awake, but gently, allowing you to function long after you'd ordinarily go to sleep. "Discovering something that could keep me functional without making me loopy was an absolute godsend," Farr said. "During the 2007 Chaos Communications Camp, I stayed up for five days on five hours of sleep thanks to Club-Mate." Even better, the beverage's proponents claim, once you stop drinking it, you can go straight to sleep if you want to.

Unfortunately, Club-Mate is mainly found in Germany and Austria, and transport to the United States is possibly only by freighter, making it difficult for American hackers to get a hold of the stuff. In 2008, hacker legend Emmanuel Goldstein imported a few pallets for a convention he was organizing, exposing many stateside hackers to the Club-Mate phenomenon.

Credit: Andrew Plumb (Community Commons)

▲ *Club-Mate is as close as it gets to an official hacker beverage.*

In 2010, Farr bought a pallet of his own and drove from the East Coast to Nevada for the DefCon conference, stopping along the way to sell Club-Mate. "Once we reached Vegas, I pretty much started selling it straight out of the Buick," Farr recalled. "Almost everyone picked up their cases within a few hours, and the last of the reserve stock I ended up trading for a place to crash."

Despite Goldstein's and Farr's attempts to evangelize their love of Club-Mate, the beverage remains hard to find in the United States. One alternative hackerspaces have found is simply to brew their own out of yerba-maté tea leaves, adding sugar and powdered caffeine to approximate the real beverage's taste and effects. To Farr, the attempts demonstrate the value of the hackerspace movement. "It's not software, it's not hardware, but it's a challenge and people have come up with very interesting results and learned a lot about making something in the process," he said. "This is what hackers do—they're confronted with an obstacle and they find a creative way around it and share what they discover."

Credit: Bre Pettis

Credit: Philip Tiefenbacher

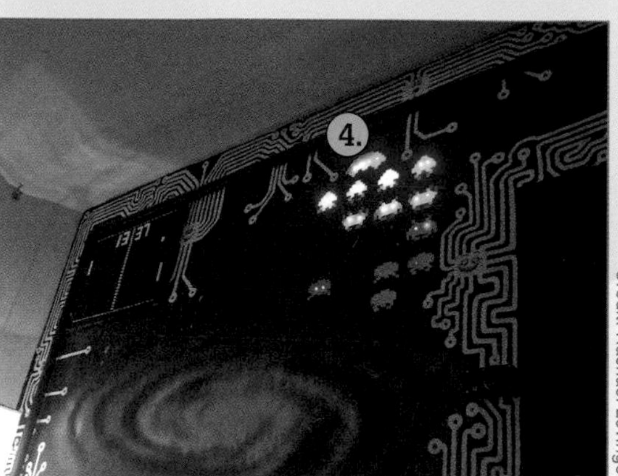

Credit: Rachael Lovinger

BUILD IT

Here's how to design your own Space Invaders LED mural.

1. Cut out alien ship shapes. You can create them yourself or research the classic shapes on the Web. The acrylic used in the ships helps diffuse the LEDs without blocking too much of their light.
2. Wire the LEDs and LED drivers. The team used 28 RGB LEDs powered by 6 LED drivers.
3. Connect and program the Arduino. Be sure to grab the TLC5940 libraries from the Metalab site if you ended up using those chips.
4. Install the acrylic spaceships in the wall and thread the wires through to the other side, where the electronics will be located.
5. Connect the mural to a power supply. Even though the Arduino can supply voltage, you'll need a separate supply to power 28 LEDs.

FURTHER READING

- Arduino code for the mural: https://whatever.metalab.at/projects/Arduino/libraries/TLC5940/examples/invaderTest/
- Philipp Tiefenbacher's TLC5940 resources: https://whatever.metalab.at/user/wizard23/tlc5940/
- The project page on Metalab's wiki: http://metalab.at/wiki/SpaceInvaders

PROJECT 13

TELEPRESENCE ROBOT

Hackers from Shanghai hackerspace XinCheJian conceived and built a robot that could help an instructor teach a class across the Internet.

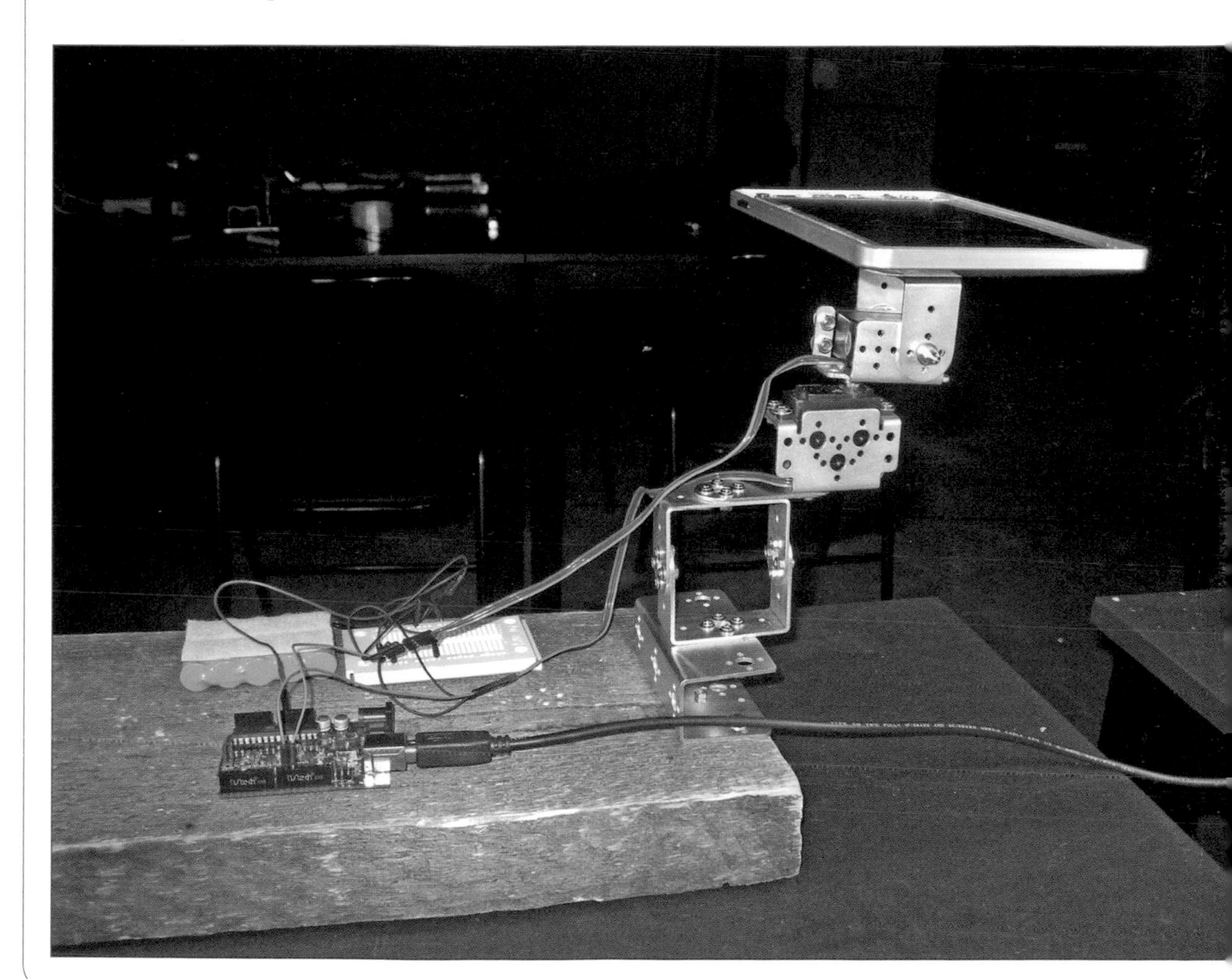

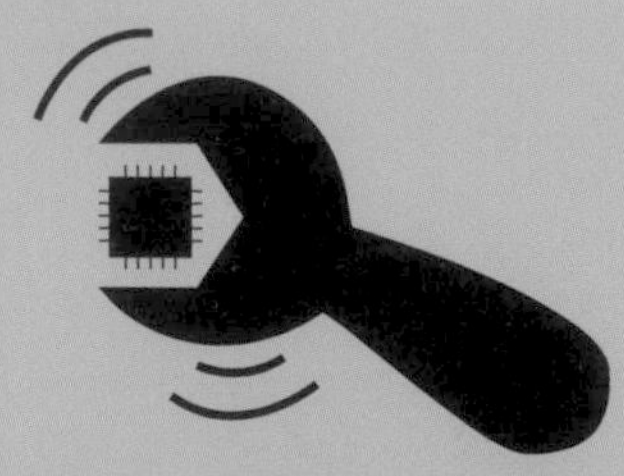

新车间

XIN CHE JIAN

http://xinchejian.com

HACKERSPACE PROFILE: XINCHEJIAN

Location:

XinCheJian

Anhua Road No.76 suite 301

Changning District, Shanghai, China 200050

http://xinchejian.com/contact-us/

Organization type: "Not-for-profit, barely legal organization," member Ricky Ng-Adam explained.

Founding date: September 2010

Number of members: 10

Dues: 500 RMB/month; "Payment in services or donations accepted."

Size of the space: 100m2

Officers/Leaders:

- Min Lin Hsieh, community organizer
- David Li, foreman
- Ricky Ng-Adam, project generator

Notable equipment: Oscilloscope, projector, MakerBot, drill, soldering irons, Dremel (a real one and a Chinese clone), and "a big pile of bamboo"

SPACE DETAILS

"We're on the third floor of an old abandoned textile factory converted into work spaces," Ng-Adam described. "The building is on a very old-style, typical Shanghai street. Our space is next to a (noisy!) dance studio. We've repainted the walls; added shelves, tables, and chairs; and cleaned up the space a bit."

Credit: Mitch Altman

▲ *The front door to XinCheJian. The poster to the left advertises one of Mitch Altman's legendary soldering classes.*

THE PROJECT

Project participants: Min Lin Hiseh, Danzel Lee, Michael Liao, Lionello Lunesu, Maksim, Ricky Ng-Adam, Swat Su, and Worlf Wang

When XinCheJian, the first hackerspace in China, heard about the Great Global Hackerspace Challenge (see sidebar), they saw an opportunity to earn some prizes for their space while working on a challenging project.

The focus of the challenge was on educational projects directed at schoolchildren. Because Shanghai, XinChJian's home city, is a hotbed of electronic and robotic innovation, the thought was to play to that strength by building some sort of educational robot. They settled on a telepresence robot, which allows an individual in a remote location to see and hear what is going on around the robot and lets him speak through it.

Credit: Ricky Ng-Adam

▲ *A XinCheJian member adjusts the hackerspace's drill press.*

Credit: Ricky Ng-Adam

▲ *Visitors and members congregate at the space.*

Credit: Ricky Ng-Adam

▲ *XinCheJian overlooks a busy commercial district.*

Credit: Ricky Ng-Adam

▲ *The telepresence robot features an interactive face that allows the bot to better connect with students.*

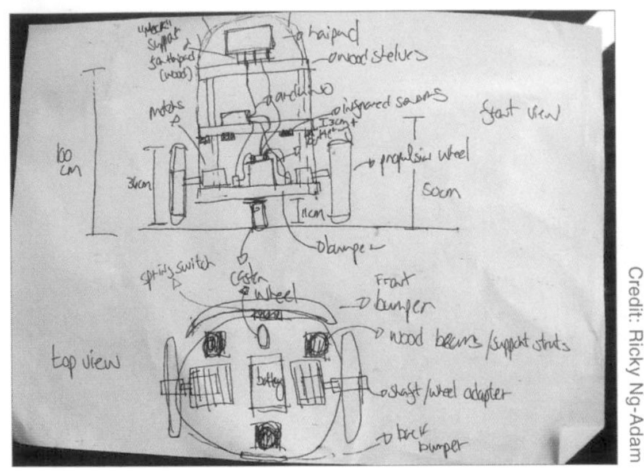

Credit: Ricky Ng-Adam

▲ *An early sketch for the robot called for a more blocky look.*

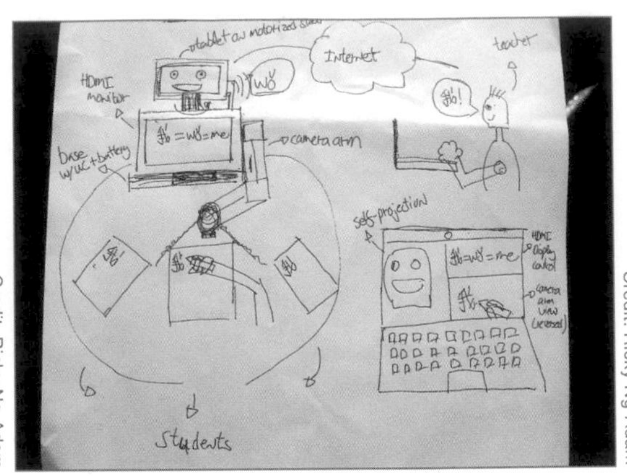

Credit: Ricky Ng-Adam

▲ *A sketch shows the robot's anthropomorphic design and an idea on how it could be controlled.*

"A telepresence robot allows teachers or experts to remotely teach classes," explained XinCheJian co-founder Ng-Adam. "It can also enable parents to remotely observe their kids' performance in class. By customizing height and appearance, we can make the robotic teacher more appealing to the students." The original plan called for the robot to roll on wheels, and the team looked to their A.R.T. Car project (see sidebar) as a starting point. However, this plan was eventually dropped as other technical problems took center stage.

While telepresence robots are available commercially, the team reasoned that they could build one more economically and increase usability by making it an open platform with schematics and code that anyone could download and alter. The fact that the team could make use of *shanzai* (knockoff) electronics from Shanghai merchants offered a temping way to complete the project under budget.

The team began by buying a *shanzai* iPad clone called a Haipad M701-R that runs Android—an easily hackable operating system released by Google. The Haipad would serve as the robot's head, sending control signals to the robot's servos, streaming video to the operator, and serving as a pleasant interactive display for the benefit of students.

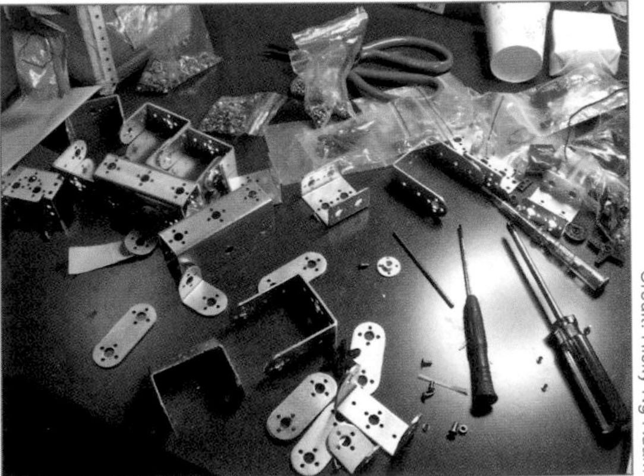

Credit: Ricky Ng-Adam

▲ *The servo-controlled neck prior to assembly.*

Credit: Ricky Ng-Adam

▲ *The completed neck awaits installation of the Haipad.*

Credit: Ricky Ng-Adam

▲ *Ricky Ng-Adam programs the Arduino.*

"The M701 in itself is an example of hacking in China," explained Ng-Adam. "Here, hacking hardware that wasn't thought to be workable together to achieve something close to a device that is much more expensive is not called a hobby but a full-time job!" However, *shanzai* products are difficult to integrate with other gadgets. "Manufacturers are in a race to produce a whole bunch and sell them before the local competitors do the same, often advertising capabilities of the underlying chipset that are actually unavailable to the user."

Credit: Ricky Ng-Adam

▲ *Team member Lionello Lunesu works on the neck.*

The team sat down and designed the robot. Because the individuals of the team were more software oriented, they had a difficult time choosing the right servos. "A real mechanical engineer would have sat down, calculated all the static and dynamic forces, [and then] selected appropriate servos for the task," Ng-Adam wrote in a blog post. "At XinCheJian, we like to start with the end and work our way back. Seriously, if you're a mechanical engineer, we'd be happy to have your help—this is very far removed from our day jobs doing software!"

The team eventually selected Tower MG995 servos. They removed the back panel of the Haipad and drilled mounting holes with a Dremel to connect the Haipad to the servo; then they added an Arduino to control the servos. Along the way, they accidentally damaged the Haipad, breaking off the touch-sensitive plate that covers the screen.

On the software end, they programmed two Android applications that would manage the communications between the teacher and robot, as well as offer a pleasant interactive personality for students. "The Android app connects to the Arduino using Bluetooth and sends commands that are then read and applied by the Arduino," the project page explained. "The user interface displays streaming video captured in real-time by the tablet computer and provides interactive controls to let the user change both the animated face expressions and the neck position." The face the team programmed into their application was designed to provide multiple expressions to help the teacher interact with the students, but they also hoped to someday add face tracking and eye tracking capability, making the expressions more truly interactive with the students.

Credit: Ricky Ng-Adam

▲ *The Haipad's read housing was removed and holes were drilled with a Dremel, allowing the tablet to be firmly attached to the servo.*

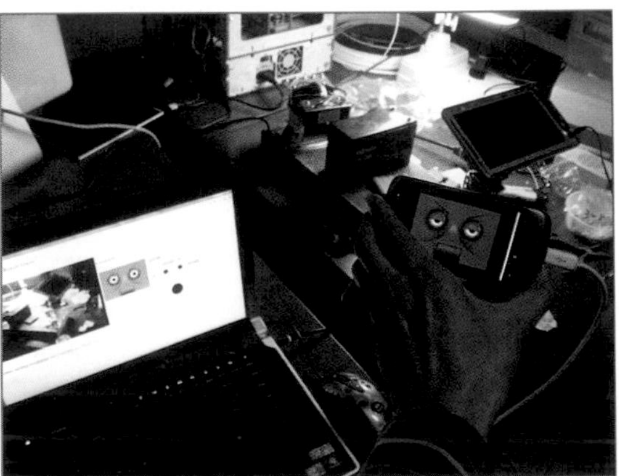

Credit: Ricky Ng-Adam

▲ *The interactive Android app—dummied up on a phone—displays the expression that school children would see on the Haipad.*

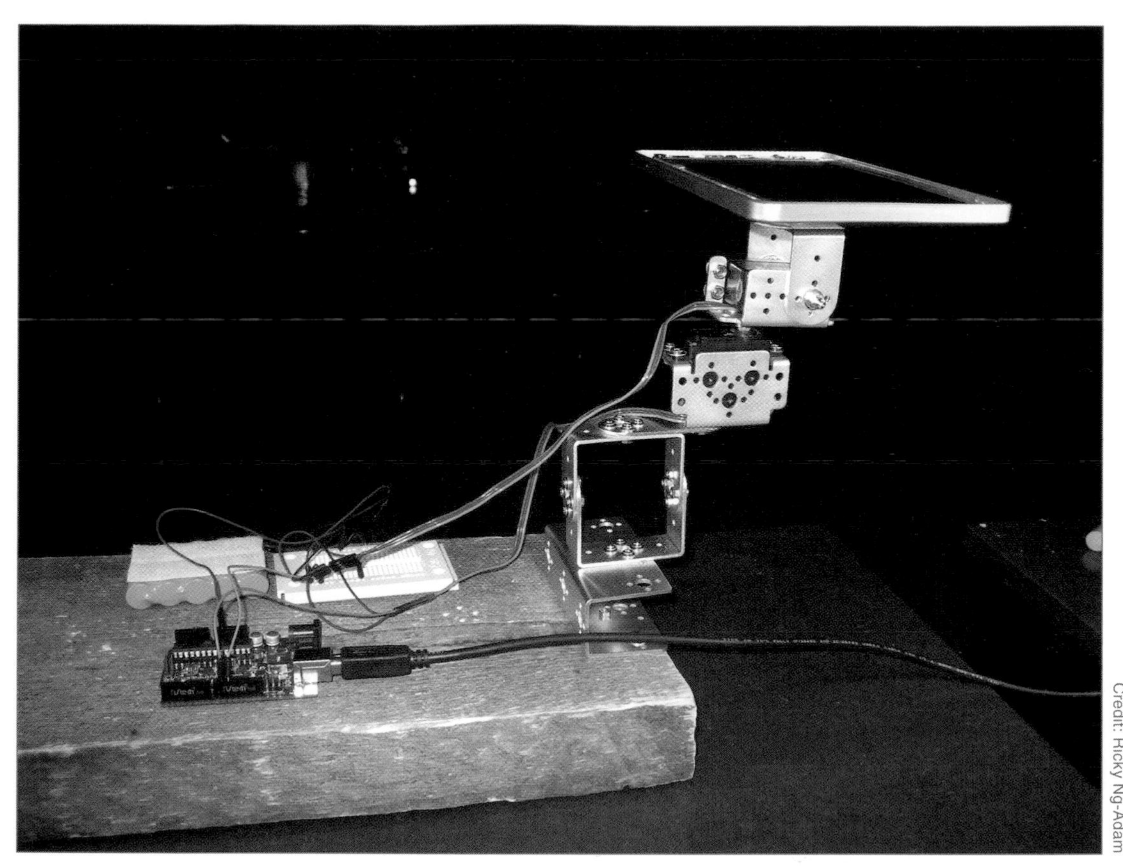

Credit: Ricky Ng-Adam

▲ *The telepresence robot as it was submitted to the contest.*

In the end, XinCheJian didn't win the Great Global Hackerspace Challenge, but they did get $900 from the contest sponsors to buy a bunch of robotics parts, and they were able to spread the word about their new space to other hackers around the world.

GREAT GLOBAL HACKERSPACE CHALLENGE

The Great Global Hackerspace Challenge was conceived by element14 (element14.com), a division of electronic component seller Newark/Farnell (newark.com) designed as a store and forum for DIYers and engineers. They hired legendary hackerspace evangelist Mitch Altman to organize and judge the contest, and around 30 spaces from around the world signed on with the promise of a $900 budget per space as well as more prizes for the winners.

Each entry had to follow a specific theme: education, interpreted in various ways by the contestants. For instance, the Hack Factory built 10:1 scale, working replicas of electronic components (described in Project 1, earlier in this book), while Pumping Station:One built a biosensor array consisting of multiple sensors interfaced through an Arduino, measuring body temperature, oxygen saturation, heart rate, galvanic skin response, respiratory volume and frequency, as well as carbon dioxide output. The stated goal of the project was to help the wearer collect his or her own data, learn how to understand it, and be able to provide medical professionals with larger data samples as needed.

Hackerspace Charlotte brought home the trophy with Feltronics (feltronics.org), a system of electronic components embedded in felt shapes designed to look like the appropriate schematic symbol.

To learn more, see http://www.element14.com/community/groups/the-great-global-hackerspace-challenge.

ALT.PROJECT

A.R.T. CAR

"RC toys are so cheap in China that it would be a crime not to give a shot at using them to make robots," the A.R.T. project page explains. A.R.T., which stands for Autonomous Robot Toy, is a robot that moves around without the benefit of human guidance, relying on infrared and ultrasonic sensors and preprogrammed instructions for keeping it out of trouble. The brain of the project is the ubiquitous Arduino, which interprets sensor data as obstacles and routes accordingly:

- Project site: http://xinchejian.com/projects/autonomous-robot-toy-a-r-t/
- Code repository: https://github.com/rngadam/ART

Credit: David Li

▲ *XinCheJian's autonomous robot was one of the first projects undertaken by the hackerspace. In this version of the project, a servo-mounted sensor scans.*

Credit: Ricky Ng-Adam

▲ *The original plan called for using the A.R.T. car as the base for the telepresence robot, but this was never realized.*

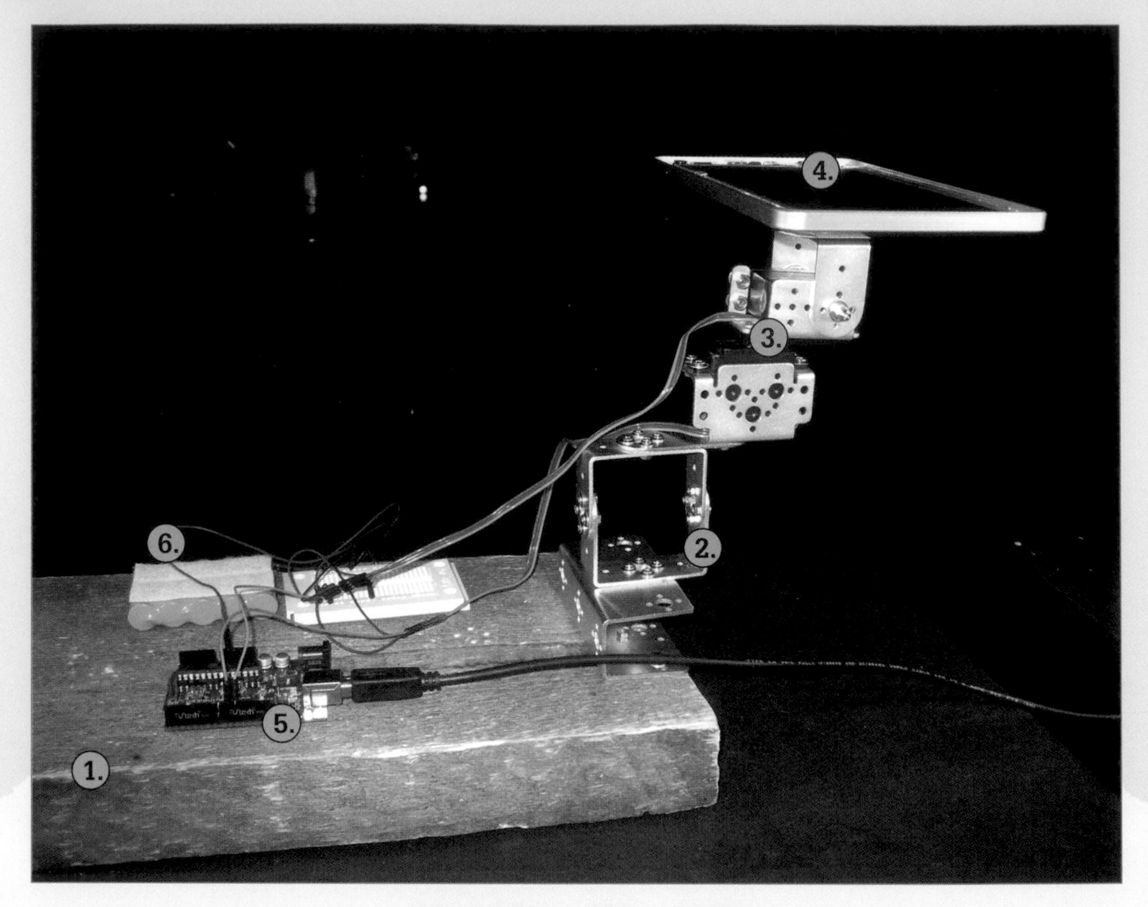

BUILD IT

XinCheJian's telepresence robot might not have won the contest, but it's still a fascinating project. Here's how you can build your own:

1. Begin with the base. XinCheJian originally hoped to mount the robot on an obstacle-avoiding autonomous vehicle.
2. Build servo neck. The team used aluminum servo hardware purchased from *shanzai* dealers in Shanghai.
3. Add servos. The Tower Pro MG995 servos the team used feature metal gears and 13kg/cm torque and moves 6° in .17 seconds.
4. Mount Haipad. The team gingerly removed the back casing of the tablet, screwed the casing to the servo hardware, and then reconnected the face.
5. Configure Arduino and Bluetooth module (the latter isn't visible in the build photos).
6. Attach the battery pack. The project team initially used a rechargeable battery pack but switched to a 12V sealed lead-acid battery.

FURTHER READING AND VIEWING

- Video summary of the project: http://www.youtube.com/watch?v=F0zPsWY_58Q
- Neck assembly summary: http://www.youtube.com/watch?v=OFtUvv18llE
- Bluetooth control of the neck: http://www.youtube.com/watch?v=2ZhRKfPWdqc
- Design notes and code repository: https://github.com/rngadam/XinCheJian-GGHC/wiki

PROJECT 14

THE POLYPLASMIC ARCOPHONE

Desiring to add a little zap to their hackerspace, members of Australia's Perth Artifactory built a Jacob's Ladder. They liked it so much they built several and set their sizzling electrical arcs to music.

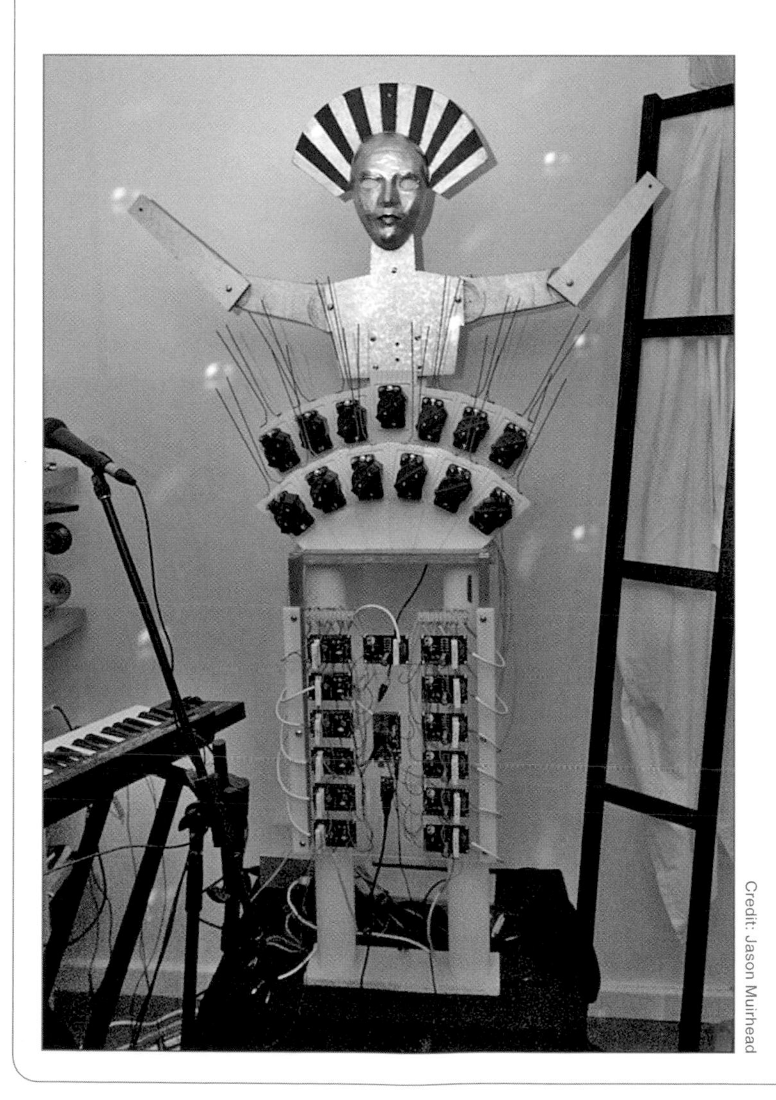

Credit: Jason Muirhead

HACKERSPACE PROFILE: THE PERTH ARTIFACTORY

Location:

Perth Artifactory

Unit 8, 16 Guthrie Street

Osborne Park, Western Australia

artifactory.org.au/

Organizational type: Incorporated association (nonprofit)

Founding date: September 2009

Number of members: 30

Dues: $100 full, $50 associate; students pay half of either level

Officers/Leaders:

- Daniel "atrophy" Harmsworth, chairman
- SKoT McDonald, treasurer
- Brendan "LordMortis" Ragan, secretary
- Peter "prd" Dreisiger, deputy treasurer
- Simon "Zignig" Kirkby, deputy secretary

Size of the space: 225 sq. meters

Notable equipment: 600mm×450mm CNC mill/router aka "Swarf-O-Mat", RepRap v2 "Mendel" 3D Printer

Space details:

"We have built in a social area made from a variety of random found furniture and office cubicle parts," Artifactory chairman Daniel Harmsworth described.

"The separation of the heavy tools has been very important given the aversion to floating clouds of atomized aluminum our members and their equipment have."

The space is very active in 3D printing and CNC technology, as well as a focus in music.

Credit: Jason Muirhead

▲ *The main room at the Perth Artifactory.*

Credit: Jason Muirhead

▲ *Artifactory members ply their soldering irons.*

Credit: Jason Muirhead

▲ *A MakerBot prints out plastic as its operator looks on intently.*

Credit: Jason Muirhead

▲ *The completed Polyplasmic Arcophone, ready to perform.*

THE PROJECT

Participants: Brett Downing, Daniel Harmsworth, SKoT McDonald, and Jason Muirhead

Deciding that their hackerspace needed something that went "zap", SKoT, Brett, and Daniel assembled a Jacob's Ladder kit they'd bought from a mail order catalog.

"We were playing with the kit when Brett came across the Arduino Tone Library and thought that we should experiment," collaborator Daniel Harmswoth recalled.

▲ *Coil drivers connected to the Freetronics TwentyTen microcontroller.*

The group decided that the kit was lacking something. "That something was the capacity to play Tchaikovsky's 1812 Overture alongside our Internet-controlled smoke ring cannon, which had been completed a week earlier."

As the sparks started popping in the hackerspace, people got interested in figuring out how to set them to music. The solution was to bypass the kit's 555, which is a simple microchip that generates a steady pulse that can be used to control a clock or other time-based project.

"A working solution was discovered by inserting an opto-coupler allowing the connection of a Freetronics TwentyTen," Harmsworth said, referring to an Arduino-compatible microcontroller.

The TwentyTen took over the timing of the electric arcs from the 555, allowing the music to trigger the Jacob's Ladder. "Shortly thereafter a hardcoded version of the *Star Wars* imperial march could be heard throughout

▲ *One of the team members tests an array of six Jacob's Ladders.*

the Artifactory," the team wrote on the hackerspace's website. However, the guys still weren't happy. The notes were weak and crackly, so it was decided that the group needed to develop its own platform to both control the music as well as the arcs of electricity. The group designed an electronic kit they called a musical coil driver.

"The coil driver is an extremely robust high inductive load driver designed to operate at audio frequencies," Harmsworth described. The driver is able to handle the 30,000 volt arcs as well as triggering them in conjunction with digital music. As a side benefit, the designers claim the musical coil driver is safer than most commercially available kits. "We put this down to a strong self-preservation instinct," Harmsworth explained.

Credit: Jason Muirhead

▲ *Electricity zaps from the prototype Jacob's Ladders. The wires were eventually replaced with more attractive ones in the finished instrument.*

Credit: Jason Muirhead

▲ *Seven ignition coils have been placed, with six more on deck.*

Credit: Jason Muirhead

▲ *The array of 13 musical coil drivers surround the Arduino Mega clone.*

In its finished form, the Polyplasmic Arcophone consists of 13 Jacob's Ladders powered by automotive ignition coils and coil driver boards. The music is controlled by an Arduino Mega clone running MIDI Processing code as well as spark-triggering software. An ATX power supply provides the juice.

The Arcophone, however, isn't merely a pile of circuit boards; the team designed a decorative rack to hold the assembly, with the thought of featuring it in live performances. It made its debut on Halloween 2010 and was featured in Euchronia, a Melbourne steampunk musical event held at year's end. "It even played 'Auld Lang Syne' at midnight on the main stage."

Simply by putting together the musical coil driver the team has learned a lot about developing and manufacturing kits, and they're looking at several other concepts to turn into commercial products.

They're not done with the Arcophone, however. They hope to add more lighting, a fancier enclosure, and the crowning touch—a gigantic executioner-style knife switch.

Credit: Jason Muirhead

▲ *The Arcophone nears completion as its decorative rack takes shape.*

Credit: Jason Muirhead

▲ *The completed Arcophone debuts.*

JACOB'S LADDER

A Jacob's Ladder, more properly known as a high-voltage traveling arc, is a device for producing a continuous series of electrical arcs that travel upward. The arc, made up of heated, ionized air, forms at the base of a pair of wires, where they are the closest, and then rises up. The wires are angled apart, so the arc of ionized air gets longer and becomes increasingly unstable. Eventually the arc breaks, only to reform at the base.

The display, which serves no practical purpose, is best known for being a prop in mad scientist movies. As one would expect, touching the arc, which carries as much as 30,000 volts, can harm a human being or ignite combustibles. It also produces ozone and nitric oxide, which can potentially harm human beings, so it's important to experiment with traveling arcs away from enclosed spaces.

Credit: Rob Flickenger (Community Commons)

▲ *Electricity crackles between the two wires of a Jacob's Ladder.*

TOOL BIN

ARDUINO MEGA

Like the classic Arduino (described in Project #1), the Mega often serves as the brains of an electronic project. Its pins receive input from sensors and switches and send signals and power to electronic components ranging from motors to LEDs, using its ATmega 2560 microcontroller to interpret sophisticated commands that can be uploaded via USB. What sets the Mega apart from its smaller cousin is its huge number of pins. It has 54 digital and 16 analog pins, allowing users to control large and complicated projects with just one board.

Credit: Mark Demers

▲ *The Arduino Mega is just like a regular Arduino, only with more input/output pins.*

Credit: Jason Muirhead

Credit: Jason Muirhead

BUILD IT

Here's how you can build this fascinating musical instrument:

1. Assemble as many musical coil drivers as you need for your design. You'll need one PCB per ignition coil. You can learn more about the kits at http://wiki.artifactory.org.au/doku.php?id=projects:musicalcoildriver.
2. Connect each coil driver to an automotive ignition coil. Typically, these drive 20kV–30kV, which is perfect for a Jacob's Ladder. The team used coils harvested from junkyard sedans, but they're looking into motorcycle coils because they are designed to operate at much higher frequencies.
3. You'll also need to attach wires to the ignition coils. These don't have to be fancy—it's common to use wires from metal clothes hangers. Intriguingly, using antennae coated with different materials changes the color of the plasma: for instance, sodium-coated antennae yield orange plasma.
4. Connect the coil drivers to your microcontroller, with each driver being controlled by a single Arduino pin. The Artifactory team used an Arduino Mega in part because of its numerous input/output pins.
5. Connect the power supply. A standard ATX power supply from a computer will do the trick. Each coil driver board has ports for power and ground.
6. Program the Arduino. You can download the code from the project page at http://wiki.artifactory.org.au/doku.php?id=projects:arcophone.
7. Assemble the instrument. The Artifactory created a humanoid design to carry the electronics, but you can make yours look any way you want.

PROJECT 15

DIY CNC ROUTER

Every hackerspace dreams of buying a CNC mill, a computer-controlled milling tool that carves designs into wood, plastic, or metal. But Harford Hackerspace didn't buy a CNC; they built their own.

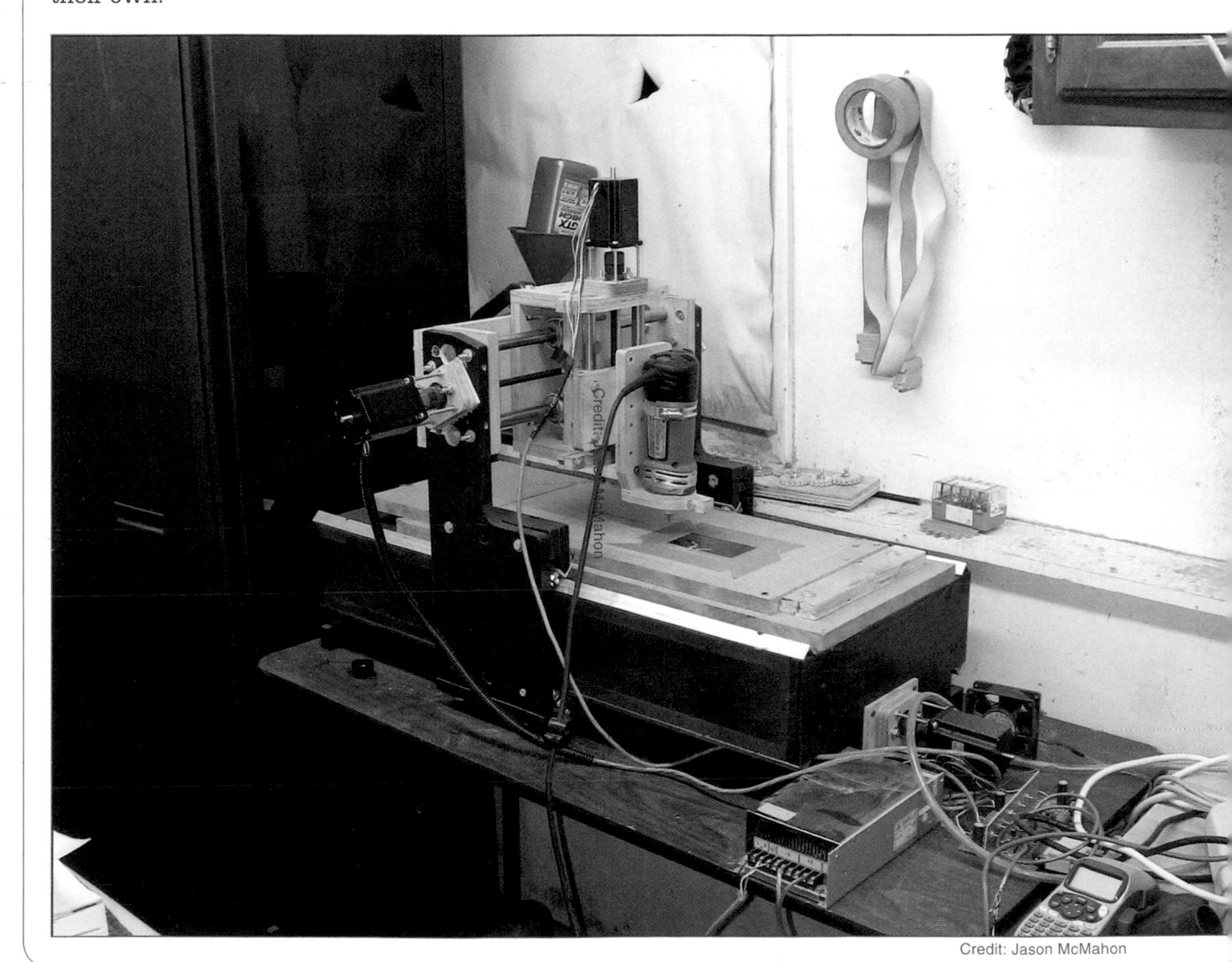

Credit: Jason McMahon

HACKERSPACE PROFILE: HARFORD HACKERSPACE

Location:

Harford Hackerspace

4602 Powell Avenue

Baltimore, MD 21206

harfordhackerspace.org

Organizational type: Nonprofit corporation

Founding date: January 2009

Number of members: 12

Dues: $50/month

Size of the space: 1,500 sq. ft.

Officers/Leaders:

- David "Squintz" Powell, president
- Jason "bsom" McMahon, secretary
- Miles Pekala, treasurer
- Paul King, board member
- Chris Cockrum, board member

Notable equipment: CNC mill, drill press, table saw, metal lathe

Credit: Jason McMahon

▲ *Harford Hackerspace members congregate for the organization's CNC User Group meeting.*

Credit: Jason McMahon

▲ *Paul King hacks with the help of his son, Laser—yes, that's really his name.*

Credit: Jason McMahon

▲ *A common sight at a hackerspace—a member happily hacking away at a project.*

SPACE DETAILS

"The space is a large two-and-a-half-car garage owned by one of our members," said president David Powell. "The hackerspace does not pay rent, which allows us to use our dues for purchasing cool tools and materials for group projects." Confusingly, the hackerspace is actually located in Baltimore. "This is because when I started the space I lived in Harford County. This has made it difficult for us to attract additional members. People from the area think we are in Harford and don't bother to check us out.

One of the projects the members have worked on as a group is an inventory system to keep track of their electronic components, junk, and other materials, so they're not buying parts they already have.

THE PROJECT

Participants: Chris Cockrum, David Powell, Paul King, Jeff Nelson, Gary Cygiel, Miles Pekala, and Jason McMahon

There are a number of expensive tools that make every hackerspace member drool: laser cutters, 3D printers, and CNC routers top the list. Intriguingly, hackerspaces sometimes decide to build their own rather than buy one of these machines. Harford Hackerspace chose to create a CNC router.

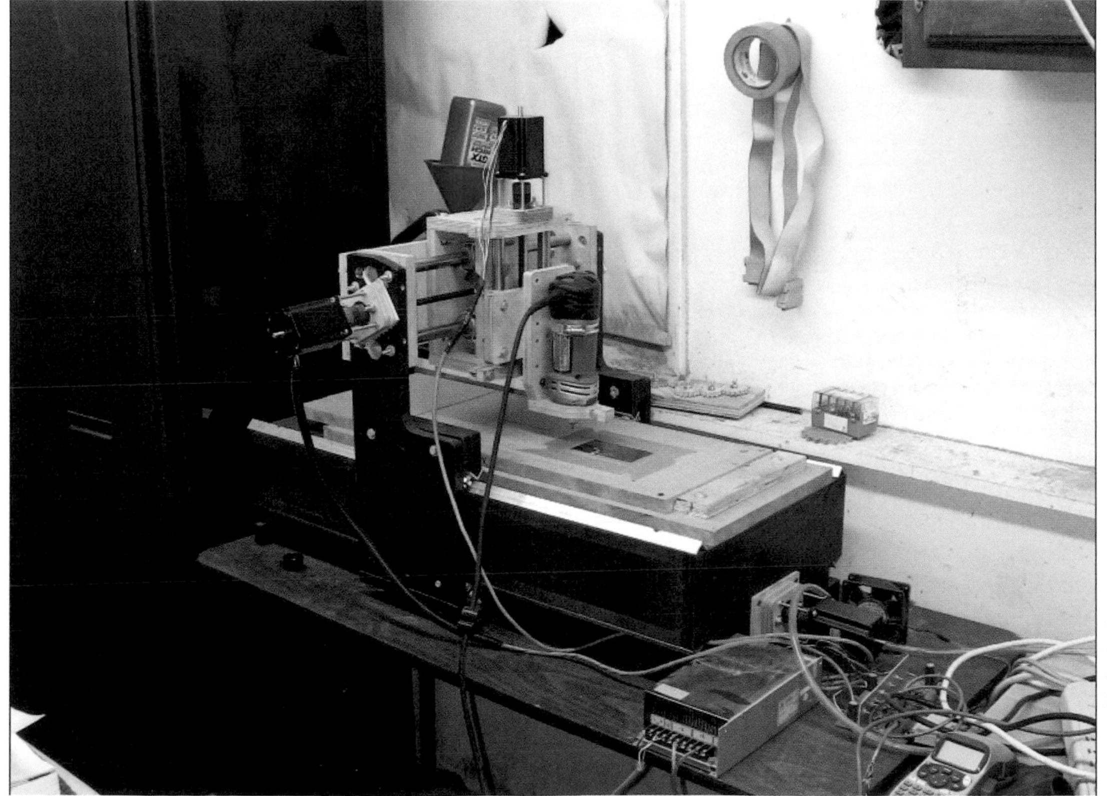

Credit: Jason McMahon

▲ *Harford Hackerspace's CNC awaits its next project.*

Credit: Jason McMahon

▲ *The CNC's powerful stepper motors are strong enough to lift a 65-pound weight.*

Credit: Jason McMahon

▲ *The gantry takes form. Eventually it will be equipped with a Rotozip rotary tool that will do the cutting work.*

"When we initially started the organization, we met at local restaurants once a month," Powell recalled. "When Paul King joined he offered to allow us to use his garage. During that same meeting, we decided that we needed a group project to get us going." Member Chris Cockrum suggested they build a CNC router.

A CNC (computer numerically controlled) router consists of a cutting tool that operates automatically, controlled by instructions sent from a computer. These instructions are created by a piece of software that converts a 3D model into a series of tool movements. The operator puts a block of material to be carved—perhaps wood, or plastic, or metal—onto the build platform and starts the machine. The tool activates, dips down, and starts carving away material. As the router works its way through the instructions, the shape begins to take form.

While a CNC router can be used to cut anything a regular power tool could, fans prize CNCs for their ability to precisely mill pulleys, gears, and other machine parts that are difficult to make by hand.

Having finally found a space, the Harford Hackerspace team got to work building the CNC, working one night a week for six months straight before they cut their first parts. Like many collaborative hackerspace projects, team members were recruited as they visited the space and expressed interest. "We simply handed them a tool and said, 'Here work on this,'" Powell said. "Chris showed a couple of us how to use SolidWorks to design the machine. Since he already had experience with building CNCs, he took a back seat and let us make our own mistakes. When we made a mistake, Chris would suggest a fix, which usually worked."

Credit: Jason McMahon

▲ *The gantry, shown close up. The horizontal bars control the x-axis and the vertical bars the z-axis.*

Credit: Matthew Forr

▲ *The CNC's controller board, running the machine's motors as instructed by a computer.*

After cruising along leisurely, the project came together very quickly. "In September 2009, the CNC came to life when we attached a pen to it and started drawing cartoon characters," the team wrote on the organization's website. "Being a bunch of teenagers trapped in adult bodies, someone decided that we should cut a throwing star from sheet aluminum." After generating the pattern with a copy of CNC software CamBam, the shape was ready to be cut. "The star took about 10 minutes to cut and was an excellent first attempt at metal production."

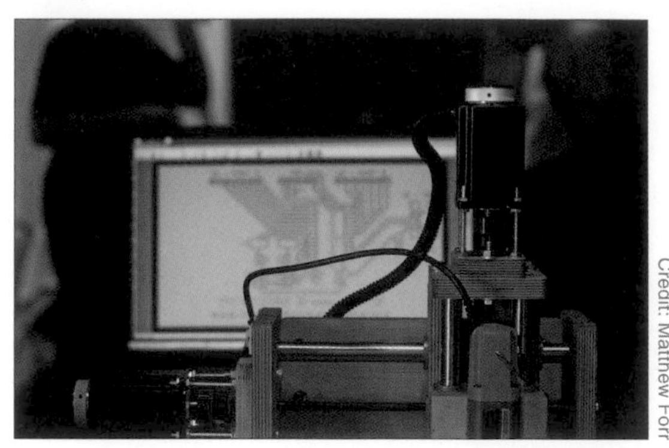
Credit: Matthew Forr

▲ *A computer screen displays the interface the operator uses to control the CNC router.*

Credit: Matthew Forr

▲ *The CNC etches a printed circuit board, foreshadowing the machine's future role at Harford Hackerspace.*

While the first draft of the CNC seemed promising, it soon proved to have some critical flaws. "We tried to make the CNC inexpensive since we had little startup money at that time," Powell said. "We used materials that wore out quickly. This required us to rebuild the machine. Then we attempted to cut materials that were too hard. This caused a little bit of damage to the Rotozip bearing."

Immediately the team began to think about ways they could improve the design. They wanted to use Baltic birch for the gantry—the original used the cheaper but also less robust medium-density fiberboard. The Y-axis bearings weren't flush and started to damage the rails, so they switched to self-lubricating bronze bearings and solid steel rails.

Ultimately, however, a second homebuilt CNC became the space's go-to machine. As for their first machine, it still operates and is fondly remembered as the group's first collaborative project. However, it awaits another retrofit before it returns to active use. "We plan to reuse the parts from our original machine to build yet another CNC, which will be much more accurate," Powell explained. "The intended purpose will be to etch PCBs."

DIY FABRICATION

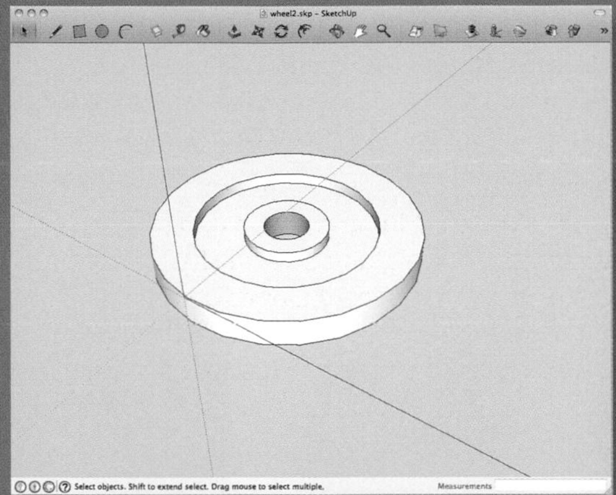

A 3D model of a wheel, generated in Google SketchUp. A plug-in allows users to export an .STL file, which ReplicatorG can convert to G-code.

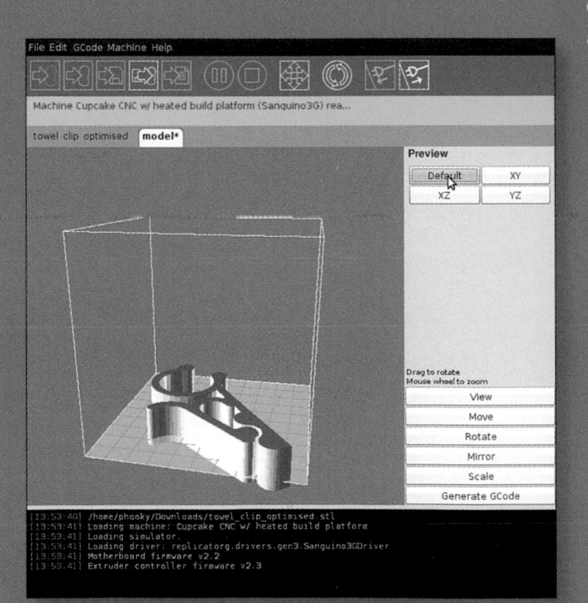

ReplicatorG previews a model. Once it's ready to go, the user clicks the button that generates the G-code.

Many hackers are obsessed with 3D fabrication, an intriguing technological movement in which individuals can essentially manufacture objects using a number of computer-controlled tools. Many hackerspaces have almost become mini factories, allowing members to go from design to prototype to small-scale production of inventions. While no one expects these spaces to replace full-sized factories, they certainly give options that once were limited to high-end prototyping shops.

The three main fabrication tools used in hackerspaces are 3D printers, laser cutters, and CNC routers. In previous chapters we've discussed 3D printers, by which plastic objects are built layer by layer. Laser cutters etch or cut acrylic and plywood with the use of a powerful laser. CNC routers use a cutting implement like a Dremel rotary tool to carve away at a block of wood, metal, or plastic.

BUILDING THE MODEL

Before the computer can tell the tool where to carve or print, it must have a model to work from. If you're looking to get into DIY fabrication, the good news is that there are numerous low-cost or even free software options to choose from.

Computer-aided design (CAD) software runs the gamut from free and simple to incredibly robust suites costing thousands of dollars and taking years to master. One of the easiest to learn is Google SketchUp (sketchup.google.com), a popular program with a vast repository of technical advice as well as legions of predesigned objects. SketchUp is available for free, or you can buy a "pro" version for $495. Another popular choice is Blender (blender.org), a free and open source package with many more capabilities and options than the free SketchUp, but with a comparably steeper learning curve.

Laser cutters don't need 3D modeling software because they work only in two dimensions. You'll still need software to design your creation, however. Popular choices include the freely downloadable InkScape (inkscape.org) as well as commercial products Adobe Illustrator CS5 (adobe.com), which runs $599, and CorelDraw Graphics Premium Suite (corel.com), which costs $699.

CONVERTING THE MODEL TO TOOLPATHS

Once you've designed your shape, you must convert it into instructions for the tool. The most popular programming language for controlling CNC tools is G-code. At its essence, G-code consists of instructions for the machine's motors. For instance, a command might tell the x-axis to move 0.5mm or draw a circle 15cm across. Other elements such as coolants and toolhead speeds are also managed through G-code. Converting the shape to G-code involves using software like ReplicatorG (replicat.org) or Skeinforge (skeinforge.com), both of which you can download for free.

Once the toolpaths have been generated, all that's left to do is print!

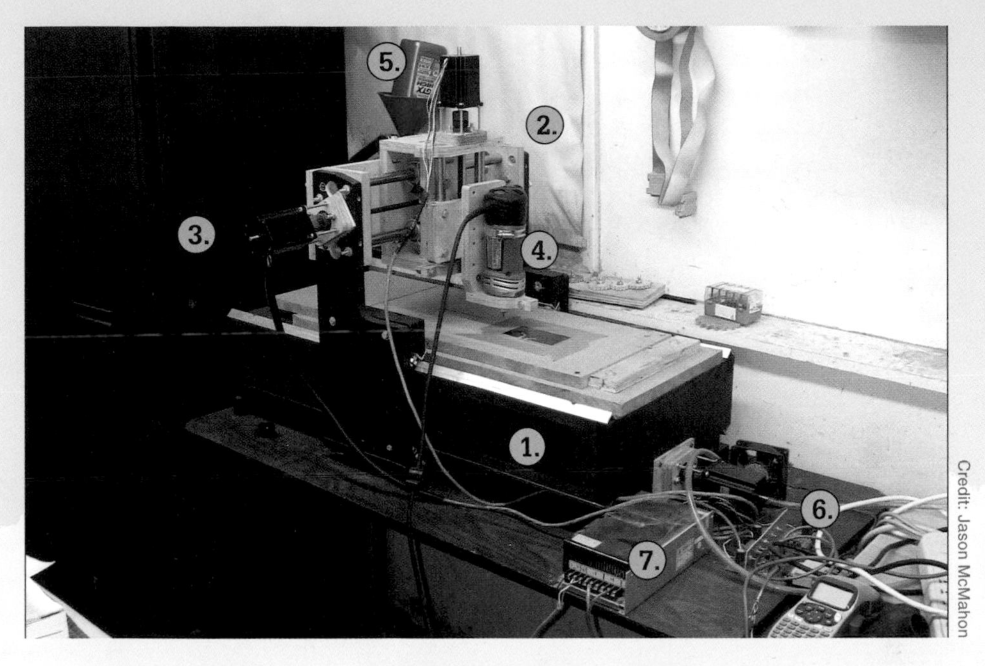

Credit: Jason McMahon

BUILD IT

Building a CNC router is an expensive and challenging project, but there are many resources out there for hobbyists and tinkerers to create their own. While there are many different possible configurations, you're sure to find a lot of advice online for helping with your particular project.

1. Build the base. While possibly the least exciting part of a CNC router, it's actually quite important because the machine's stability is very important for ensuring accurate outputs.
2. Build the gantry. This assembly travels along the machine's y-axis and controls the toolhead's movement along the x- and z-axes.
3. Attach the motors. Harford Hackerspace used 23-305-DS8A stepper motors that are rated for 425 ounces of pull-in torque. You'll need three, one for each axis.
4. Add the routing tool. Many hobbyists use a rotary tool such as a Dremel, but an actual contractor-grade router is a possibility as well.
5. Add lubrication. Lubrication will increase the life of your rails. Simply take a quart of oil and give the moving parts a squirt when things get tight.
6. Add the CNC controller, which interfaces the motors and the computer.
7. Connect the power supply. The hackerspace created a custom supply. The one you need will depend on many factors such as the draw of your motors. cnczone.com's forums contain many suggestions for helping you choose the right one.
8. Connect the computer. The hackerspace used a Dell GX280, but pretty much any nonobsolete machine will work.

FURTHER READING

- BuildYourOwnCNC, an invaluable how-to site that the Harford hackers referenced during the build:
  http://buildyourcnc.com/default.aspx
- CNC Zone is another site for DIYers interested in building CNCs:
  http://www.cnczone.com/
- Harford Hackerspace's wiki article about their CNC project:
  http://wiki.harfordhackerspace.org/index.php?title=Cncproject

PROJECT 16

LED MATRIX GAMING SYSTEM

Members of Swedish hackerspace Forskningsavdelningen built an LED matrix that can play simple videogames and use the Internet to connect with opponents around the world.

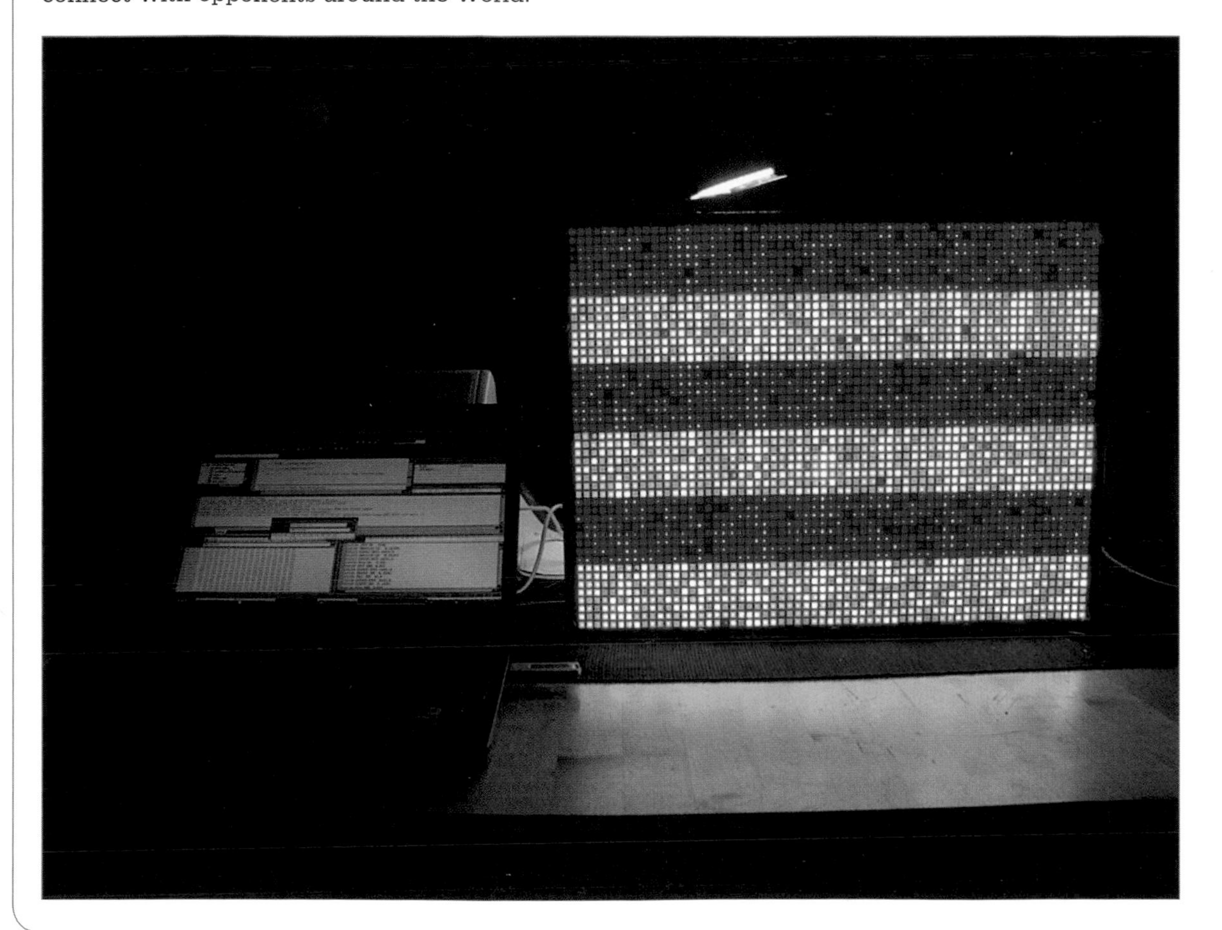

HACKERSPACE PROFILE: FORSKNINGS-AVDELNINGEN

Location:

Forskningsavdelningen
Norra Grängesbersgatan 26
SE-21450 Malmö Sweden
forskningsavd.se

Organization type: "Nonprofit would be the best description," member Davey Taylor described. "However, since we want everyone to be able to focus on what they love and not be bothered with bureaucracy in any way, we have opted to not turn Forskningsavdelningen into a registered legal entity."

Founding date: August 2008

Number of members: "Undetermined since we do not require membership and do not keep lists," Taylor said. "Best guess: 20–30."

Dues: "None; we currently do not require membership in any form. Anyone is welcome, at any time. We hope to be able to keep it this way."

Size of the space: 400 sq. m., shared with another organization

Officers/Leaders: "None," Taylor said. "We have a completely flat hierarchy. Call it 'anarchy' if you will, but we prefer to avoid the word due to its political associations."

Notable equipment: "Because of our economic structure, Forskningsavdelningen has only basic tools and hardware but varied enough to allow for almost any kind of work by numerous people simultaneously. People are happy to bring more advanced equipment when necessary."

SPACE DETAILS

"Forskningsavdelningen interests me primarily because of the creative environment it provides," Taylor explained, "and the fact that you are always surrounded by competent people who are willing to help out at a second's notice if you run into a problem." The hackerspace's members have extremely varied interests and come from very different backgrounds, according to Taylor.

"I've used tools and components in ways I didn't think possible for me," Taylor said. "This has really promoted the whole 'thinking outside of the box' mindset. Hacking is often described as 'using things in ways they were not meant to be used,' and you are often forced to do just that, which can sound a bit cumbersome. But people always admit that when they look back at their projects, that's one of the things they were most happy about because of what they've learned through this way of working."

Credit: Andie Nordgren

▲ *The main hacking area at Forskningsavdelningen.*

Credit: Peter Rukavina

▲ *Forskningsavdelningen is located in a building full of leftist organizations. One of them serves free vegan meals every Monday.*

Credit: David Tolnem

▲ *Forskningsavdelningen's storage shelves are weighed down by electronic components, manuals, tools, and junk.*

THE PROJECT

Participants: Davey Taylor, Boomlinde, Faidros, Jimbo, Jenx, JonasB, Kepyla, Limpkin, Mackt, Olleolleolle, Salkinitzor, Stuge

Many hackerspace projects begin in one direction and then evolve into a different project altogether. Davey Taylor and a couple of friends from Malmö, Sweden's Forskningsavdelningen hackerspace built Reggaebox, a Caribbean-themed music player intended for an outdoor music festival in Denmark. One of the Reggaebox's features is a small dot matrix display on the side of the box that scrolls text while the music plays. Taylor and his friends liked the matrix so much, they decided to enlarge it and make it the focus of the project. The thought was, what if they made the dot matrix display bigger, large enough to play videogames on? They decided to build an 18-bit color, 64×48-pixel LED matrix add-on with game controllers and exchangeable game cartridges.

They began by building the display out of 48 smaller LED modules, each consisting of 8 rows of 8 LEDs, for a total of over 3,000 full-color pixels. Each LED module has its own LED driver, a chip that manages the data flow that allows each LED's brightness and color to be controlled individually. Each column of 6 LED modules has its own data channel, making for a much faster display than if all the drivers were serially connected.

"The first phase was focused mainly on getting a general idea of the layout and finding the best components for the drivers," Taylor described.

▲ *The completed matrix gets a software tweak via a connected laptop.*

▲ *The matrix is made up of 48 smaller LED matrices, each packing 64 RGB LEDs.*

▲ *The project's laser-cut frame awaits the LED drivers and other electronics.*

▲ *The Forskningsavdelningen team etches a circuit board for the project.*

With some 3,000 pixels, the display needed its own sound and graphics hardware, as well as a Zilog microprocessor to run the whole thing. "The original idea was to drive the panel from the CPU," Taylor explained. "But after doing the math mid-project, this turned out to be completely unfeasible so I had to learn FPGA programming."

FPGA, which stands for Field-Programmable Gate Array, is a semiconductor chip for managing large numbers of circuits. The team found a Cyclone II FPGA board on eBay and developed custom code for managing sound and graphics.

"The board also contains a 512Kb SRAM used for sample data," Taylor said. "This replaces the original DRAM that came with the board."

The FPGA board is hooked up to the Zilog CPU that manages the serial communications with the LED drivers. As a bonus, the Zilog board has an Ethernet port, allowing it to be connected to the Internet. In addition to this Ethernet port, the project has game ports for attaching controllers, MIDI ports, and an SD card reader.

Credit: Davey Taylor

▲ *A team member solders up a connection.*

One of the biggest hurdles the Forskningsavdelningen team encountered was power management. Thanks to the project's genesis as an add-on to a boom box, the team always intended for the matrix to run off battery power.

"I had to find a lot more insight into high-power switching components and DC-DC converters—none of which I had any previous experience," Taylor explained.

Another challenge was to create the circuit boards for all the custom hardware installed in the matrix. Taylor laid out the boards using PCB design software DipTrace.

"The first iteration of the design was intended for manual production," Taylor said. "We tried different production solutions but ended up getting the best results using toner transfer from non-glossy, laser-printed paper run through a laminator together with the substrate. Aligning the two sides of the design was tedious.... We managed to get 0.2mm traces etched nicely. After a few tries, we got the hang of soldering 0.25mm pitch components manually onto the PCB." (For more information on etching your own PCBs, see page 85 in Project #7.)

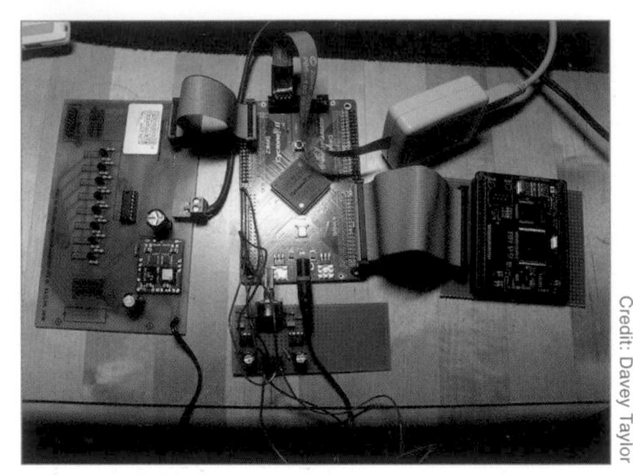

Credit: Davey Taylor

▲ *The matrix's circuit boards, prior to installation.*

Credit: Davey Taylor

▲ *The matrix comes together as the 48 LED drivers are connected to each other.*

For the driver boards, the team elected to go with a professional PCB service. "We ended up using PCB-CART because of the low price even with solder mask and screen print," Taylor explained. "Quality was okay, but two of the boards did have manufacturing errors."

This was the first time Taylor had ever ordered professionally produced PCB boards, and PCB-CART's automated format. When ordering PCB boards created by a third party, you upload your layout files and hope for the best. This means that Taylor had to fully master the DipTrace's export features.

"We decided the best thing would be to export layers in discrete files," Taylor explained, "rather than have all the layers collected together in a single file. This worked out well, except the two previously mentioned production errors, and I was very happy with the price, delivery times, and final results."

After he got the boards, the team tried to solder the SMD (surface-mount) components by hand. These are minuscule resistors, capacitors, and other electronic parts often placed by machine due to their tiny sizes.

"This was too cumbersome, so I got a reflow oven from China and learned about SMD production," said Taylor. (See the Tool Bin sidebar for more information on SMD components.)

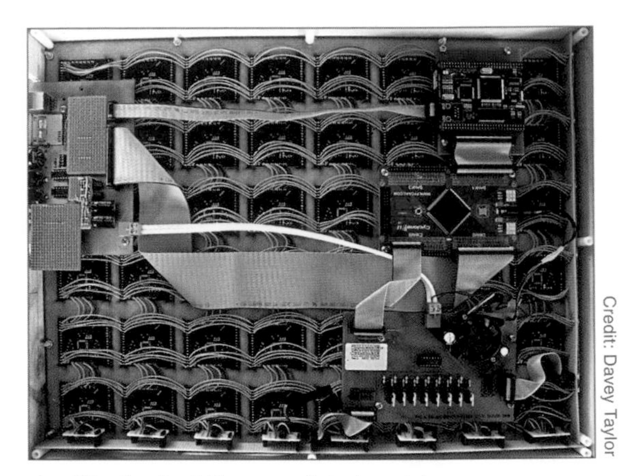

Credit: Davey Taylor

▲ *The back of the completed matrix.*

After about four months, the matrix was functional. Taylor then used it to play a game of Pong across the Internet with a Swiss hacker named Limpkin, who had his own LED matrix of a completely different architecture than Forskningsavdelningen's.

The team has published the project as open source, and all schematics, PCB designs, and instructions are available on the project site.

"I have not yet had the time to publish the code publicly, but it is freely available upon request," Taylor said.

TOOL BIN

SURFACE-MOUNT COMPONENTS

As technology progresses, electronic components get smaller and smaller. Traditional through-hole electronics include resistors, capacitors, and LEDs that have wire leads that go through the PCB where they are soldered in place. However, surface-mount components (SMD) are much smaller and mount on the surface of the board. They're so minuscule that electronics manufacturers typically use machines to place the components and solder them in.

Despite the challenges, many electronics hobbyists and hackerspace members have attempted to master the techniques necessary to use these components in their projects, reasoning that SMDs will one day be the standard electronics components and through-hole will ultimately go away.

"Surface mounting can be a bit tricky, and there are several methods recommended by different people across the Web," Taylor said. "We tried a few of them and found that adding generous amounts of flux to a row of legs, then adding solder to the tip of a fine soldering iron prior to sliding it quickly over the row worked best. Usually we had to remove a few bridges afterwards, but all in all it worked very well. A flux dispenser is absolutely necessary for SMD work."

Credit: Jeff Keyzer (Community Commons)

▲ *Surface-mount components are compact, are inexpensive, and cannot be easily soldered by hand. They're also the future of electronics.*

When attempting to solder in numerous components, hackers use special ovens that heat up large amounts of solder so an entire board can be soldered in one treatment. "The method I used was to pre-solder all of the pads, followed by dispensing a sticky flux and attaching the components to the flux before putting them in the oven," Taylor said. "The oven can hold six boards and takes 5–10 minutes or so for each pass. It's a very useful machine, highly recommended for small-scale work." Many hackerspaces repurpose commercial toaster ovens for their reflow, being sure to mark them as not being safe for food use after lead-based solder goes into them.

FORSKNINGSAVDELNINGEN POLICE RAID

Some people who hear the term *hacker* immediately associate it with computer crime, and the Malmö police department can probably be included in that group.

It all started with a factual, albeit breathless, piece in Malmö newspaper *Sydsvenksan* describing a visit to Forskningsavdelningen and the activities of its members. The article described how the members of Forskningsavdelningen learned about picking locks (a commonplace hackerspace activity called *locksport*) and also talked about technology. Reading the translated piece, it seems evident that some elements, as relayed by an outsider, might raise the hackles of a conservative police official, such as the locksport angle or the fact that some of those interviewed didn't want to use their real names.

A week later, the police raided the entire building that Forskningsavdelningen shares with several left-wing organizations. Their stated goal was busting Malmö Hardcore, a punk rock concert promoter allegedly selling alcohol without a license at a concert they were holding in Utkanten (Edge).

"They began by clearing everyone out of the building, having them leave their belongings behind," described Taylor. Oddly, the police reportedly brought a full computer forensic team with them. "It's been argued that it seems very improbable they would bring a whole team of technicians to a night club," Taylor observed, "especially since they went directly for our space while the minority of 'regular' officers searched the rest of the very large building."

Among other things they confiscated were three laptops, four desktops, and locksport paraphernalia including two key cutters and a number of key blanks. When the key copying machines were found, the police accused the space of planning crimes. "As far as I can tell, it is not illegal to own a key grinder or key materials." Taylor said. "The grinder was used as a teaching aid for new members, allowing them to make their own personal key for the space as a fun exercise free of charge."

The following year the police agreed to release all the computers, but the locksport tools were never returned. No charges were filed. "All in all, though, people are happy about the raid since it gave us more PR than we could ever have imagined," Taylor said.

To see Forskningsavdelningen's post about the raid, visit http://forskningsavd.se/2009/11/29/i-can-haz-moar-bout-teh-reid/.

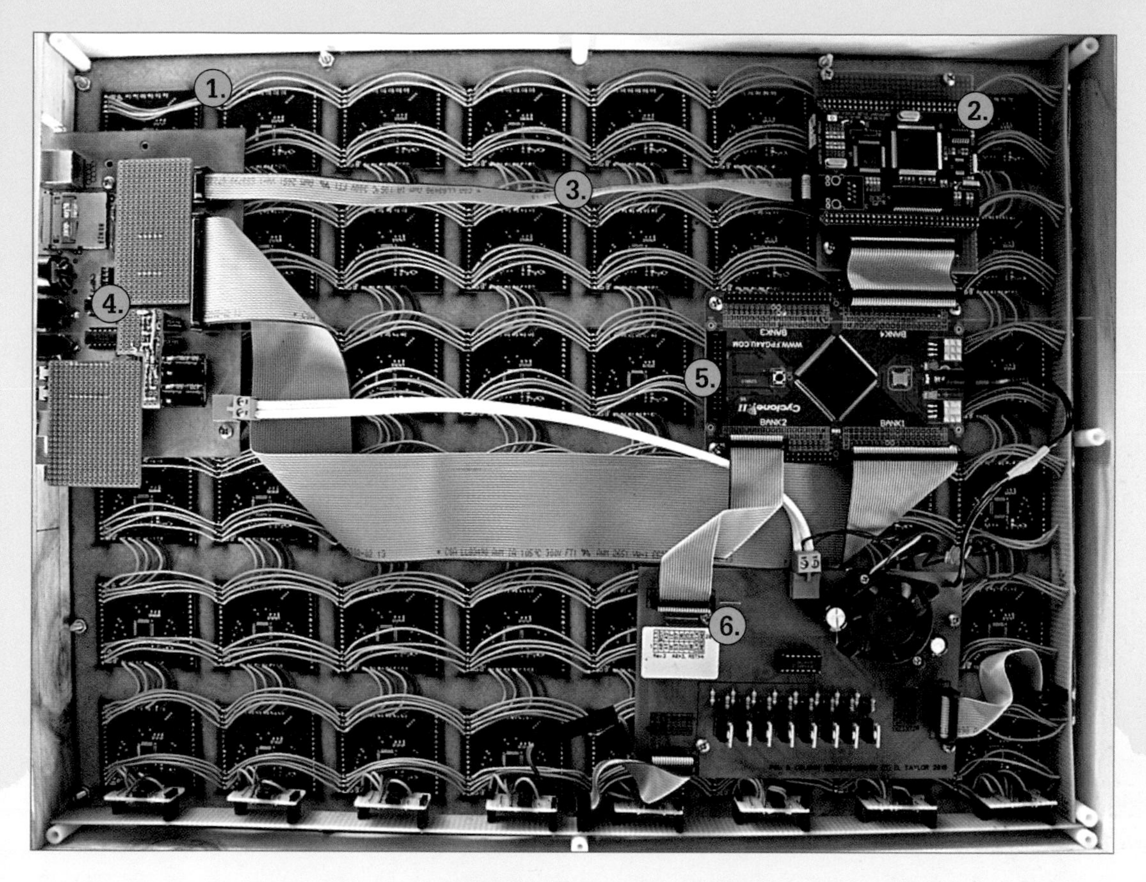

BUILD IT

Forskningsavdelningen's LED matrix is practically a computer in itself, able to run simple games off of an SD card. Wanna make your own and challenge the Swedes to a game of Pong? Just follow these steps:

1. Laser-cut the enclosure out of wood. This provides a framework for positioning the various circuit boards.
2. Install the microprocessor. The hackerspace used a Zilog eZ80F91 CPU.
3. Connect the STP24DP05BTR LED drivers. Each driver board manages 64 LEDs.
4. Assemble the I/O boards. These manage the DC, Ethernet, MIDI, SD card, and other ports on the side of the unit.
5. Add the FPGA board. The FPGA board manages the matrix's audio and video.
6. Connect the power board. The power board contains a cooling fan, as well as resistors and capacitors that manage the power.

ADDITIONAL RESOURCES

The project site:

http://forskningsavd.se/wiki/index.php/PET-GCM

Watch a video of the Pong match between Davey Taylor and Limpkin:

http://www.youtube.com/watch?v=Kk-d30FdG_0

PROJECT 17

MAME CABINET

Yearning for the era of arcade games, the members of theTransistor hackerspace built a multiple arcade machine emulator (MAME) to play the classic games whenever they want.

HACKERSPACE PROFILE: THETRANSISTOR

Location:

theTransistor

560 S 100 W

Provo, UT 84601

thetransistor.com

Organizational type: "Just a club"

Founding date: October 2009

Number of members: 5

Dues: $90/month regular membership, $35/month limited membership

Size of the space: 1,500 sq. ft. plus storage

Officers/Leaders: Founding members:

- Tim Anderson
- John Fenley
- Deven Fore
- Gordon Cooper
- Erik Lowry

Notable equipment: MIG Welder, PCB etching station, 24" vinyl cutter, auto-opening heat press, chop saws, miter saws, circuit analyzers, a couple 100MHz scopes, kitchen

SPACE DETAILS

With such a small number of members, the space is more personalized than many hackerspaces. For instance, every member has his or her own desk. "We move everything around a lot," founding member Deven Fore explained. "It is getting to be a little more stable, but from time to time we rearrange the entire space." Open nights often feature 10 or more visitors, leading to such silliness as epic Rock Band karaoke.

"The space was never intended to make money or be revolutionary," Fore said. "We all have day jobs. This is what we love to do. Just hang out, play some video games, and occasionally etch a new circuit for something or other. We all believe that information should be free."

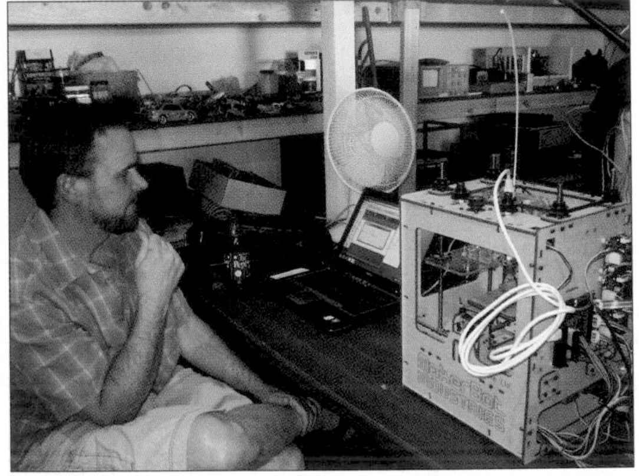

Credit: Deven Fore

▲ *The classic hackerspace accoutrement, the MakerBot 3D printer, gets some work.*

Credit: Deven Fore

▲ *theTransistor's open night makes for a full house.*

Credit: Deven Fore

▲ *Members and visitors congregate for a Dungeons and Dragons game.*

THE PROJECT

Participants: Tim Anderson and Deven Fore with help from Dave Armstrong, Gordon Cooper, John Fenley, Max Gudmundson, Erik Lowry, and John Wharff

The hackers at theTransistor had pretty much every video game console available, but something was missing. They wanted to recapture the glory of the classic arcade machine. "A MAME cabinet was the logical choice," recalled Fore. A MAME cabinet resembles an upright arcade machine, but rather than having a single game, it contains a full-fledged computer in the cabinet that has many different classic video games loaded on it.

Credit: Deven Fore

▲ *theTransistor's MAME cabinet after completion.*

Credit: Deven Fore

▲ *While much of the cabinet was routed on a CNC machine, some of the finishing work was done in-house.*

Credit: Deven Fore

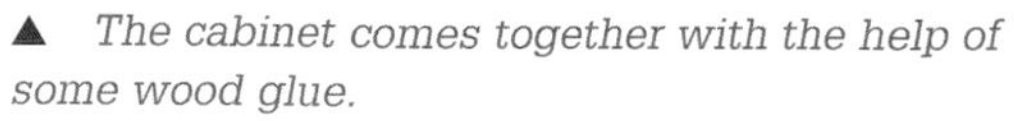
▲ *The cabinet comes together with the help of some wood glue.*

Credit: Deven Fore

▲ *Sound is supplied with garden variety computer speakers situated inside the cabinet.*

theTransistor's members divvied up the roles: Fore designed the basic look of the machine and tackled the software end, while Tim Anderson designed the cabinet in Google SketchUp. Over the next couple of months, the first version of the cabinet took shape.

As with many large carpentry projects, getting everything to look good on paper was far easier than actually making things physically fit together. "The version one plans were created without tolerances, and a few other small errors," Fore said. "This took some extra woodworking to make everything fit as we wanted."

There were more snafus. The plans called for the wrong side of the control panel to be routed.

"To solve this, we clamped it to the bottom of the control panel and used a router to route it as a template," said Fore. "This created the correct top for the control panel, and the insides were still routed correctly." Another problem was that the first cabinet was built too big to fit through some doors. Eventually a second cabinet, identical except for a few tweaks, was built.

The group ordered joysticks, buttons, and a control board for interfacing the controllers with the computer. These accessories are considered critical for having an authentic arcade experience, as they closely resemble the equipment used on classic machines. Plus, with multiple sets of buttons on the MAME's control panel, two, three, or even four people can play simultaneously, with the control board interpreting as the players hammer on the buttons.

Credit: Deven Fore

▲ *The access panel in the back of the cabinet opens to reveal a familiar site—an ordinary PC.*

Credit: Deven Fore

▲ *The underside of the control panel shows the buttons and joysticks wired to the controller board.*

Credit: Deven Fore

▲ *A visitor tries out theTransistor's MAME.*

The cabinet and control panel are merely cosmetic to make the MAME resemble an arcade machine of yore. The heart of the MAME system is an ordinary PC running specialized software. The first thing the group had to choose was the front end. This is the program that presents a menu of all the possible games to play. "Originally, we wanted to use Puppy Arcade," Fore explained. When that didn't work out, they considered writing the interface themselves. "Then we found Maximus Arcade, which runs on Windows XP. About one to two hours of tweaking, and it was working exactly as we wanted."

After a couple of months, the MAME project was done. After the high-fives the group gathered around and played some video games.

MAME

Credit: Marty McGuire

▲ *Members of Pittsburgh, PA, hackerspace HackPGH play on the space's MAME, recalling the heady days of arcade machines.*

MAME stands for multiple arcade machine emulator. It's basically is a computer that runs a program able to interpret the original arcade game programs (called ROMs) as well as those of video games running on long-obsolete consoles like the Atari 2600 and Nintendo 64.

Arcade machines have long been associated with the hacking community, with pivotal scenes in such classic hacker movies as *War Games*, *TRON*, and *Hackers* set in arcades. Many hackers who grew up on such movies—or who wasted whole afternoons in arcades—yearn to experience the scene again. However, buying a classic arcade game machine is an expensive prospect. Not only do these vintage devices cost a lot, repairing a 25-year-old electronic gadget isn't cheap either. Add the fact that each machine can play only one game, and it makes buying one of these old machines simply not worth it.

The modern MAME provides the arcade experience with the advantage of being able to play dozens or even hundreds of games, coupled with the convenience and economy of running on modern hardware and software. Win!

TOOL BIN

GOOGLE SKETCHUP

▲ *An exploded view of the MAME cabinet using Google SketchUp.*

▲ *A partially exploded view of the side of the MAME cabinet.*

▲ *A Google SketchUp sketch of the rear of the MAME cabinet.*

SketchUp is a 3D modeling program created by Google and intended as an alternative to commercial design programs that often cost hundreds or thousands of dollars. By contrast, the basic version of SketchUp is free, while the pro version costs $495 for a single user license. SketchUp can be used by architects and engineers as well as filmmakers, game developers, and—you guessed it—hackerspaces! While SketchUp is considerably easier to use than some related AutoCAD programs, it still has a steep learning curve.

The program has been used to build virtual models of everything from house plants to fully equipped sports stadiums. Even better, Google's 3D Warehouse includes freely downloadable models contributed by users.

In the case of theTransistor's MAME cabinet, the creators used SketchUp as a virtual sandbox to design each component and see how they fit together when assembled, saving a lot of tinkering time.

To learn more about Google SketchUp, see Que's *Using Google SketchUp 8* by Bud E. Smith.

BUILD IT

If you're pining for the glory days of arcade games and like the efficiency of being able to play dozens of games on one machine, a MAME is for you!

1. Choose your design. You can find multiple MAME cabinet designs on the Web, you can create your own, or you can simply use theTransistor's design. If you choose the latter, you can download the schematics from http://thetransistor.com/projects/mame/.
2. Output the cabinet parts. theTransistor used a CNC machine, but very careful scroll saw and band saw work will also suffice.
3. Attach joysticks and buttons. You can buy these on the Internet fairly easily, though sometimes they can cost a lot.
4. Install t-molding. This is the cosmetic bumper that covers up the bare wood at the edges of the cabinetry.

5. Mount and wire the controller board. theTransistor used an I-Pac4, a PCB that costs about $65 and accepts input from up to 56 controllers.
6. Install the monitor and speakers. These are normal PC components, and any models you have lying around will probably work, as long as they can interface with your computer of choice. The hackerspace used a 28" LCD monitor.
7. Install the marquee—the sign at the top of the cabinet. You can get your design printed at most print shops for a few bucks. theTransistor used a fluorescent light from a hardware store to backlight the design.
8. Add the computer. Most modern PCs can be used for MAMEs, as long as they can handle the front-end software (theTransistor used a program called Maximus Arcade) and the game ROMs. Note: There are many pirated ROMs floating around the Internet. For legal copies, find an Internet vendor who displays the appropriate licenses on their website.
9. Load the software. In addition to the computer's regular OS (most use Windows XP), you'll need the front-end software—the MAME software (which interprets the classic arcade software). You can also add emulators able to run game console software, like old NES and Sega Genesis games.
10. Waste a lot of time playing video games!

PROJECT 18

BOOK SCANNER

Members of Oakland's Ace Monster Toys (AMT) have too many books. The solution? Digitizing them with this homebrewed book scanner.

HACKERSPACE PROFILE: ACE MONSTER TOYS

Location:

Ace Monster Toys

6050 Lowell Street, #214

Oakland, CA 94608

acemonstertoys.org

Organization type: California nonprofit

Founding date: June 2010

Number of members: 25

Dues: $80; $40 "starving hacker"

Size of the space: 1,600 sq. ft.

Officers/Leaders:

- Al Jigen Billings, president, board member
- David Rorex, treasurer, board member
- Robbie Trencheny, secretary
- Stefan Hristu, board member
- Christian Fernandez, board member
- Cliff Biffle, board member

Notable equipment: 80W laser cutter, table saw, wood lathe, chop saw, band saw, drill press, metal saw, two MakerBots, various soldering irons, Zen Toolworks CNC mill

SPACE DETAILS

"The space is in a reconditioned warehouse that is used by a number of small businesses in North Oakland," described president Al Jigen Billings. "We have an 800 sq. ft. upstairs where our members can hang out, coding or doing electronics work and sometimes co-working. The downstairs is our woodshop and our small metal shop area with our laser cutter.

"Our lower floor has triple-phase 220 power and concrete floors with light wells from the roof, and the upstairs is a well-lit large room. We've found that it mostly meets our needs. We've primarily done three things to customize the space. We've extended our 220 power in the downstairs to make it more available. Our members have installed a ventilation system from the shop areas to allow gases from the wood and metal shop areas (and the laser) to vent to the outside of the building. We had to install a larger door in the metal shop in order to wheel our 750 lb. laser into the space—it was literally too wide for the doorway."

Credit: Al Jigen Billings

▲ *The modest exterior of Ace Monster Toys.*

THE PROJECT

Participants: Al Jigen Billings, Daniel Reetz, and Myles Conley

Sometimes a hackerspace begins a project out of whimsy; other times it is pure necessity that drives it. "I own around 12,000 or more books but have been trying to downsize for years," Ace Monster Toys president Al Jigen Billings described. "I have a garage filled with boxes of books and have to deal with shelf after shelf whenever I move." The thought that followed was, we live in a digital world, so he should be able to digitize his books.

Credit: Al Jigen Billings

▲ *The hackerspace's upstairs hacking area lets members code or solder away from the noise and dust of the wood and metal shops.*

Credit: Al Jigen Billings

▲ *AMT members play around with RC handsets and Kinect motion trackers.*

Credit: Al Jigen Billings

▲ *The AMT gang shows off their projects at Maker Faire Bay Area.*

Credit: Al Jigen Billings

▲ *Version one of the book scanner featured a heavy wooden frame and overhead light.*

Credit: Al Jigen Billings

▲ *The cradle, which holds the book in an open position, is made of MDF and scrap wood.*

Credit: Al Jigen Billings

▲ *Adjusting the cradle, which will hold the book in an open position.*

Some book scanners involve ripping the pages out of the binding and running them through a sheet-fed scanner, but Billings wanted a system that would leave the book undamaged. "Daniel [Reetz] came up with the DIYBook scanner (diybook scanner.org) a few years ago during an Instructables contest. It has been very successful and has spawned a pretty active community. Rather than reinventing the wheel, we decided to iterate on his design. After that, we actually became friends with Daniel and he has recently moved to the Bay Area, so it worked out to be rather fortuitous."

The basic design of the DIYBook scanner consists of a cradle that holds the book, a glass plate that holds the pages in place, mounts for holding inexpensive point-and-shoot cameras, and lights for illuminating the pages. When digitizing the book, a person opens the volume to the first page to be imaged, places the book in the cradle, lowers the platen, and then triggers the cameras. When all the pages are imaged—the project website claims to be able to scan a 1,200-page book in one hour—the images are loaded into a software utility that sorts, de-skews, and gathers the images into eBook form.

Credit: Al Jigen Billings

▲ *Version one of the scanner was overbuilt with thick wood and large metal angle braces.*

Credit: Al Jigen Billings

▲ *The platen keeps the book at the perfect angle for photography.*

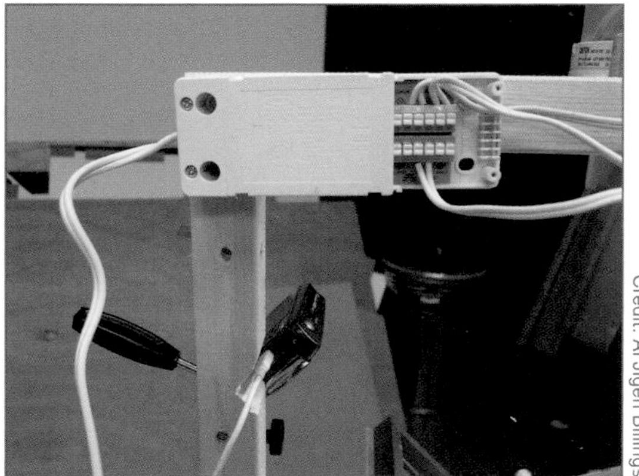
Credit: Al Jigen Billings

▲ *The lighting system was ultimately ditched in favor of clip-on shop lights.*

One thing missing from the project is a mechanism for turning the pages manually. "The reason that we don't have a page-turning mechanism is that it is a hard problem to solve well," Billings said. "It is difficult to come up with a solution that will work well on any sized book, books of any type of paper, and that are guaranteed to be reliable and not harm the book. The human finger is much more reliable. The difference between a page-turning scanner and a non-page-turning one is a lot of time and hassle so we—and the DIYBook scanner community—have largely stayed away from that."

When AMT settled on the book scanner project, they began construction using MDF cut with shop tools. In doing so, however, they discovered just how tricky manual carpentry can be. "One of the problems Al and I ran into was not being able to cut wood to precise enough angles," team member Myles Conley recalled. "While I ended up buying a chop saw—which made everything go much faster—it was obvious that wasn't the solution for most people."

This first version was overbuilt and heavy, and worse, they had a hard time getting the platen aligned properly. "Daniel and I tried a few prototypes, including the one where we accidentally glued the glass on at the wrong angle and broke it trying to fix things," Conley said.

The solution to both the clunkiness of the MDF design and the platen difficulties was to create a frame that could be designed on the computer and output on the hackerspace's laser cutter, allowing for more accurate measurements and tighter tolerances.

"The switch from MDF to laser-cut plywood was an inspiration on Myles's part," recalled Billings. "Lots of people in the DIYBook scanner community have built scanners. Everyone looks at each other's designs and builds one based on the inspiration of others. It turns out that getting the platen just right is the hardest part for most people. Myles realized that if we came up with a good laser-cut design for the platen, that it would make everything much easier for others." The second iteration offers an easier build process and a much snugger platen fit.

Another difficulty the team experienced was finding the right solution for lighting. This is an important aspect of the scanner because it impacts image quality. Not only must the cameras' sensors have plenty of light to take proper pictures, but the platen also can pick up glare from other light sources, so dimming the ambient lighting and shining a light directly on the platen makes for the best images. After building and wiring up a large overhead light for the MDF version of the scanner, they abandoned them for the laser-cut version.

Credit: Al Jigen Billings

▲ *The completed book scanner receives last-minute tweaks.*

Credit: Al Jigen Billings

▲ *The laser-cut parts of version two show the lighter construction of the new design.*

Credit: Al Jigen Billings

▲ *The new platen awaits installation in the book scanner.*

Credit: Al Jigen Billings

▲ *Version two of the scanner features a laser-cut cradle and is not able to accommodate large books.*

The reason for this turnaround was the genesis of the laser-cut scanner as a kit, the idea being that the design files could be distributed on thingiverse.com or the cut pieces sold through the mail. "Lights drive the cost of a kit up substantially," Conley explained. "Mark 2 is designed to be short enough that a normal swing arm lamp can work above the page. This does lead to problems with wobble, which we haven't solved yet. That may be an inherent compromise in keeping the whole device small and lightweight."

In addition to distributing the laser cutter patterns freely, the team has decided to sell beta kits at cost. "The overall goal is to have a flat pack of parts that, when combined with a few standard items from Home Depot, would allow anyone to build an easy book scanner."

INSTRUCTABLES

DIY website Instructables.com consists exclusively of step-by-step instructions on everything from building furniture to fixing your computer to learning a new magic trick. Even areas that fall outside of traditional DIY topics are covered, like relationship advice, organizational skills, and travel suggestions. If you want to know how to do something difficult, chances are there's an Instructable for it.

▲ *Instructables shows how to do it. Do what exactly? Practically everything.*

BUILD IT

Building your own book scanner can be a simple proposition, especially if you have access to a laser cutter and Ace Monster Toys' design files. This Build It section assumes you're building version one of AMT's scanner:

1. Build the base. This supports the platen from below, as well as a hinge on which the platen swivels.
2. Build the platen and connect it to the hinge on the base. Note the handle in the front of the platen, making for rapid page changes.
3. Build the surrounding frame, which will support the cameras and lights.
4. Add camera mounts and cameras, positioned to look down at the pages.
5. Add lights. For version two, the AMT team ended up using a shop light positioned over the platen.
6. Load book scanning software on your computer to organize and concatenate your scans.

FURTHER READING

Project website with information on the laser-cut version of the scanner:

- http://wiki.acemonstertoys.org/Book_Scanner

Book scanning software:

- http://scantailor.sourceforge.net/
- http://sourceforge.net/projects/bookscanwizard/

PROJECT 19

OPENDUINO

Luxembourg hackerspace syn$_2$cat wanted to build a system that would alert members and visitors when the space was open. The resulting setup uses a web interface to open the space and deactivate the alarm.

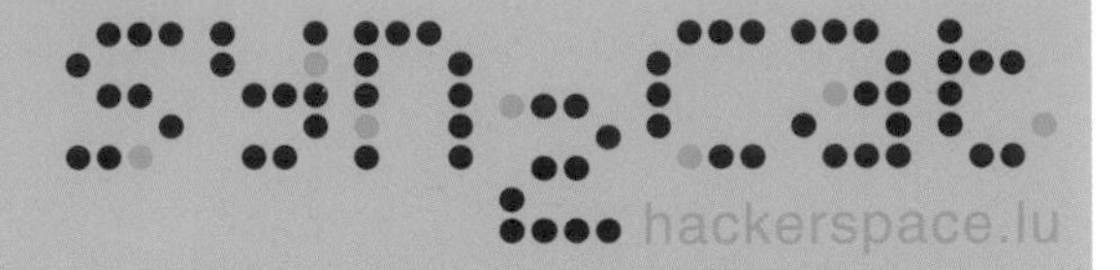

HACKERSPACE PROFILE: SYN_2CAT

Location:

syn$_2$cat
11, rue du cimetiére (am Hueflach)
L-8018 Strassen
Luxembourg
www.hackerspace.lu

Founding date: February 3, 2009

Number of members: 41

Dues: 114€/year. "We suggest to our members to pay as much as they can afford, taking into consideration their employment status, age, as well as how much they value the hackerspace."

Size of the space: 120 square meters

Officers/Leaders:

- Marc Teusch, secretary general
- Jan Guth, deputy secretary
- Georges Toth, treasurer
- Charel Büchler
- Michel Hoffmann
- Georges Kesseler
- Marc Tholl

Notable equipment: Soldering table, MakerBot, video projectors, library, computer museum, server rack, cocktail bar

SPACE DETAILS

syn$_2$cat shares space in an old school building with another organization, making for cheap rent but limiting their space. "We're allowed to occupy two of the four rooms," member Georges Kesseler described. "Both rooms still feature the chalkboard from its school days. We designated one room as the workshop/lab room, while the other one is our chill-out room where most of our meetings take place."

The organization has equipped its chill-out room, also called the "ChillyChill," with sofas, chairs, and a light-up coffee table. "It also features a self-built bar holding a three-month supply of Club-Mate and a kitchen with two fridges and three microwave ovens," Kesseler said. "The lab room has a dedicated corner for soldering work. People's lockers are also located there as well as many cupboards with electronic stuff and tables to work on. A nice book library holds a lot of technical knowledge to read through."

Credit: Prometeus

▲ *The hackerspace's main lounge, called the ChillyChill. Even the coffee table has been modified with floor lighting.*

▲ *syn$_2$cat's main hacking area, filled with computers and hackers.*

▲ *Hackerspace members enjoy Tschunk cocktails on the patio.*

▲ *syn$_2$cat's computer museum with OpenDuino visible on the left. Note the circular window; the hackerspace has rigged a light-up sign showing the space's street number, 11.*

THE PROJECT

Participants: Tschew, Kwisatz, Gunstick, sim0n

Every hackerspace has a need to control entry to the facility, with reasons ranging from protecting expensive equipment from theft to restricting entry to paid members. In the case of syn$_2$cat, their lease limited them to five radio frequency identification (RFID) badges, meaning that not every member could turn off the alarm. The solution was to build a system that would use one of the RFID badges to deactivate the security system, with a computer interface to manage it.

"You come to the front door, open your Wifi-enabled device, and open a webpage. There you enter your username and password into a webform," project lead Gunstick described. "Then the alarm turns magically off. At the same moment, the front door LEDs start flashing to indicate that you can now unlock the door with the doorknob. Inside, you can open the two rooms with a normal physical key, which every member has." Thirty seconds after the alarm is deactivated, the website is updated showing that the space is open.

Initially, OpenDuino was simply a pushbutton that changed the space's status on the website, allowing members to know when they could be assured of entry. However, that didn't solve the problem of members not being able to gain entry whenever they wanted.

"I had the idea to add functionality so that one alarm badge could be used by everyone," project member Gunstick said. The only way to turn off the alarm is to wave the RFID badge in front of the reader. "The main idea is to not change the basic working of the building's alarm system, but to add the functionality on top."

▲ *Want to visit syn$_2$cat? Be sure to check out their website to see if the space is open.*

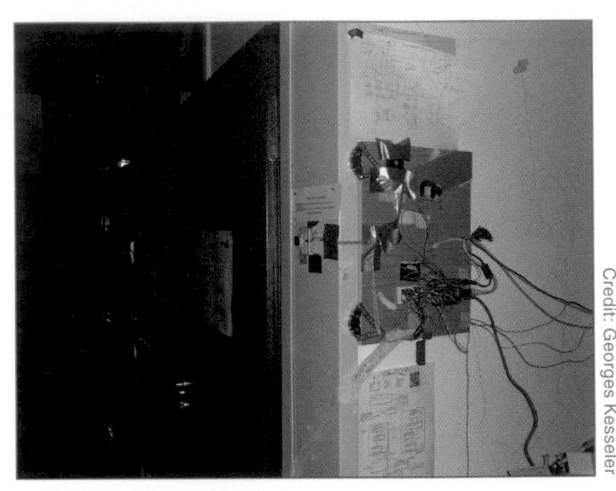
Credit: Georges Kesseler

▲ *OpenDuino at home by the front door.*

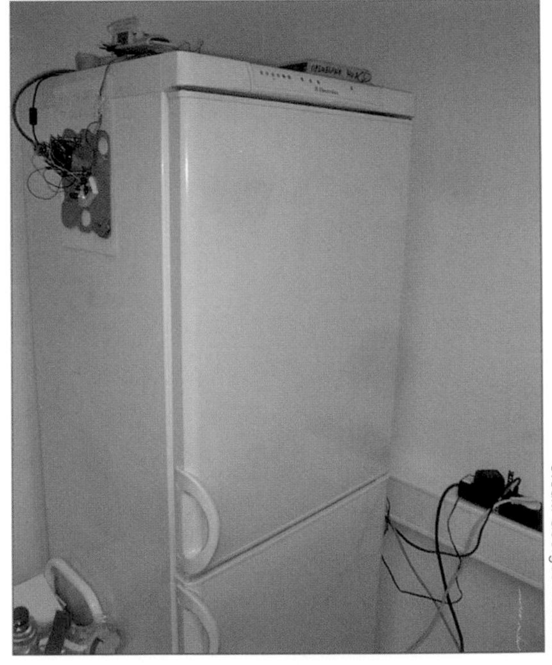
Credit: Georges Kesseler

▲ *OpenDuino was once attached to a fridge by magnets.*

Credit: Gecrges Kesseler

▲ *The RFID tag (at right) on its server-controlled arm. Ultimately, the arm was deemed unreliable and was swapped out for reed switches.*

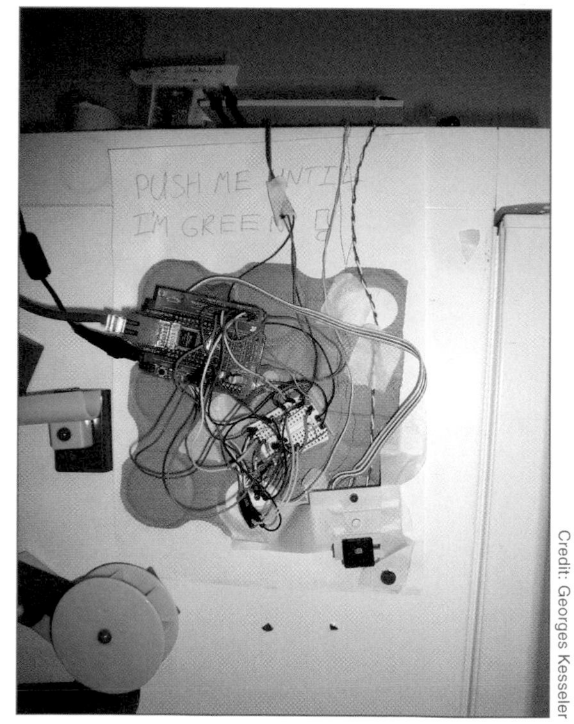

Credit: Georges Kesseler

▲ *OpenDuino's physical interface consists of a black "Windows" button at the bottom right of the assembly.*

How could the group allow additional members access without changing the system or upsetting its landlord? "RFID is radio waves," Gunstick explained. "I did some tests to see if the signal could be proxied via a long cable with coils at each end. I put the badge into one coil, and when I held the coil on the other end of the cable next to the reader the alarm would toggle. Solution!"

An early version of the system called for the RFID badge to be affixed to a servo. "The badge is fixed on the end of an arm, which the servo brings in contact with the coil," Gunstick explained. "As the first version already used an Arduino and Ethernet shield, I just continued with that and connected the servo."

One disadvantage of that version was that members still had to push the button to change the status on the Web. Logically, thought Gunstick, that status should update automatically. "I constructed custom switches located inside the locks of the lab and chill rooms. They detect if the lock is closed or open. Depending on this, the space is set to open or close status."

Even better, the space installed an electronic door lock to offset the price of cutting keys for every new member. Now, OpenDuino could control the locks as well as the alarm. "Instead of purchasing the manufacturer's expensive PC interface, I soldered two wires to the button of one of the door lock remotes."

Because the project's Arduino pinouts had been maxed out, the group tried tripping a switch with the servo's arm. "That was so unreliable that finally the servo was abandoned and replaced by reed relays," Gunstick said. "I think there is no other way to use a relay to switch high-frequency, low-voltage RFID signals."

The project has seen numerous revisions since its inception, and the team had to battle bugs with a series of kludges. "Right from the beginning there were problems with the Ethernet shield," syn$_2$cat member Kwisatz recalled. "On a 1Gb switch it works for some time and then stops responding. To debug, Gunstick put a 10Mb hub on the line. And with the hub, it always works. So until today there's the 'OpenDuino hub' doing its work of adapting good old 10Mb to 1Gb."

Another issue created by using a coil to transmit the RFID data is that the reader couldn't detect whether the badge was staying in front of the reader. "The signal is too weak and the reader interprets it as continuously waving the badge back and forth," Kwisatz said. "The result is that while the badge is on the coil, the alarm stays in a continuous on/off cycle. This has been resolved by taping a photo detector onto the alarm status LED. As soon as the alarm has the right status, the badge is removed."

Of course, the hardware was only part of the equation. On the software end, OpenDuino needed to be able to access a database of members and their logins and passwords. Kwisatz took the lead on the PHP development on the backend. The project experienced almost continual development, going from a single script to more than 20 different files. "We first started off by authenticating our members against a custom user group in the mediawiki database," Kwisatz said. "We have now arrived at a modular design where any backend can serve as authentication, as long as it adheres to an authentication interface."

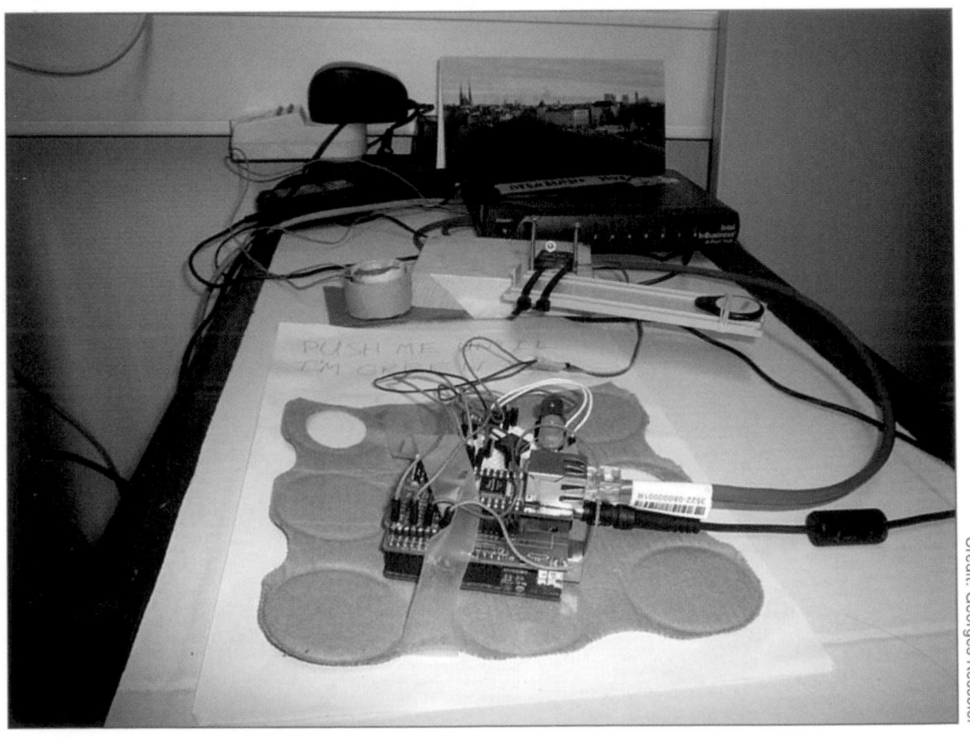

Credit: Georges Kesseler

▲ *The Arduino and its Ethernet shield interfaces between the server and the alarm system.*

The code has evolved so that its notification feature is just one of a number of plug-ins. For instance, the Garbage plug-in checks a calendar for trash and recycling pick-ups and sends an SMS to remind a designated member.

The team hasn't stopped and is already working on OpenDuino 2. "The main difference is that a lot of code and logic will be in a dedicated PC, which will have batteries," Kwisatz said. "This way, the solution also works during power cuts and Internet downtimes."

"Another evolution will be to have more intelligence built in," Gunstick added. "OpenDuino should be more aware of the various states and act accordingly. Currently, when the alarm activation fails, nothing happens. The new code will have checks and retry loops for those cases." There's even discussion of using the system to automatically control lights and blinds in the space and programming an Android app to serve as the interface.

TSCHUNK

In Project #12 we learned about Club-Mate, the caffeinated beverage of choice among European hackerspaces—and by extension, those found in the rest of the world. As with any soft drink, people have been experimenting with alcoholic versions. The most successful of these is Tschunk, which originated in a Berlin nightclub called Club Forschung ("Club Research"), where the drink came to be called "Club For-Chunk", or eventually, Tschunk.

The cocktail is superficially similar to the Brazilian national drink, the Caipirinha, which is prepared by muddling sugar and lime chunks at the bottom of a highball glass, filling the glass with crushed ice, and then topping it off with booze. With Tschunk, you pour dark rum and Club-Mate over the ice.

Fans love the Tschunk for the way the alcohol seems to accelerate the effects of the caffeine; it falls into the same niche as the rum and coke.

Credit: John Baichtal

▲ *Club-Mate may be the unofficial soft drink of the hackerspace movement, so it's only natural that hackers turned it into a cocktail.*

SYN$_2$CAT'S COMPUTER MUSEUM

Credit: Georges Kesseler (CC-SA)

▲ *syn$_2$cat's computer museum is packed with old hardware in various states of functionality.*

Credit: Georges Kesseler (CC-SA)

▲ *A Commodore 64 and Apple II flanked by a Commodore PET and a Mac Classic. An Ericsson Mini PC is visible at the bottom right.*

Many hackerspaces become a dumping ground for old computers. For the most part, it stems from an honest desire to share theoretically useable equipment. syn$_2$cat, however, has turned their lode of legacy technology into a museum. They have a Commodore 64, a Vax Station 3100, a Commodore PET with two dead keys, an Apple II in need of a new power supply, a Mac Classic, three Atari STs, an Atari Portfolio, an Ericcson Mini-PC running Windows CE 1.0, a NeXT Colorstation Turbo, and a lot more.

Check out: https://www.hackerspace.lu/wiki/ComputerMuseum

TOOL BIN

RADIO FREQUENCY IDENTIFICATION

RFID tags have been in use for years by shopkeepers and warehouses as a way of keeping track of property. The technology consists of a tiny tag with a coiled antenna built in. Often the tag has no power supply; instead, it gains just enough energy from the radio waves of the reader to transmit a code.

Hackers, however, have more fascinating uses for the technology than locating pallets. They've used RFID tags to track the location of robots or to enable one component to 'know' when another is close by. One group built a musical instrument that could be 'programmed' by placing RFID-embedded tiles in a row, and a sensor detects which tags are present and plays music accordingly.

Many people understand the technology enough to have concerns about RFID devices being used to violate the privacy of ordinary citizens. For instance, tags have become so cheap—as low as $.05— and small (tiny enough to glue to an ant) that hackers fear that governments and corporations will conceal them everywhere, building up databases of people's movements and habits, or that poorly secured data will be accessible on the tags. Accordingly, privacy advocates have developed ways of blocking RFID signals, ranging from spoofers that overlay one code with another to enclosing a tag in a tinfoil wallet to prevent unauthorized scanning.

Credit: kentkb.com

▲ *An RFID tag; its silvery antenna coil interfaces with the tag reader.*

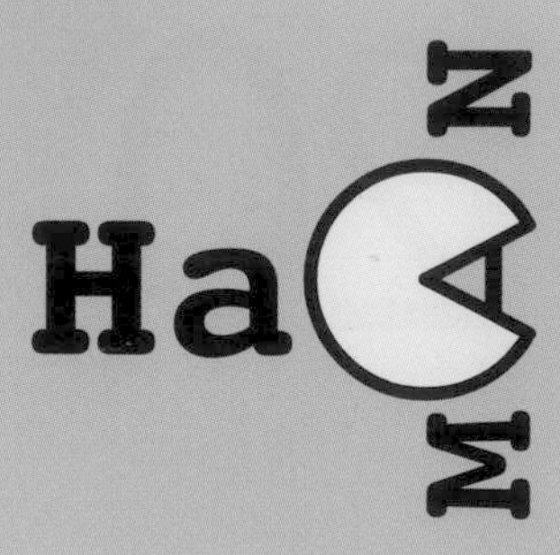

HACKERSPACE PROFILE: HAC:MANCHESTER

Location:

HAC:Manchester

36-40 Edge Street

Manchester, M4 1HN

UK

http://hacman.org.uk/

Organizational type: Unincorporated

Founding date: June 2009

Number of members: "We don't have any official membership structure," Bob Clough described. "There are about 10 central attendees and a number of people who drop in for our hack sessions."

Size of the space: "3,000 sq. ft., spread over two floors," Clough said. "The workshop area at the back of the ground floor is around 300 sq. ft."

Officers/Leaders: "Most of the necessary organizational work is done by myself and Paul Plowman," Clough said.

Notable equipment: "We've mainly been focusing on electronics and have oscilloscopes, soldering stations, and a bubble etch tank free for people to use. We do also have a small number of machine and hand tools, pillar drill, jigsaw." Members are working on building a RepRap Huxley, a 3D printer.

SPACE DETAILS

HAC:Manchester (or just HAC:Man for short) gets their workspace donated by a larger organization, the Manchester Digital Laboratory, also known as MadLab.

"The building is a 200-year-old former shop and therefore has a lot of quirks," Clough described. "None of the walls are straight! We have been at MadLab since it opened in January 2010 and have been helping out with the fit-out. We have assisted in building walls and cupboards, fitting a new kitchen, and redecorating the entire bottom floor. Sometimes we think the straightest walls in the place are the ones we've built." MadLab is a technology-focused community center, and they let HAC:Man use the space in exchange for maintenance work and publicity. "We have been to various events including Maker Faire UK and FutureEverything's Handmade, taking our projects and massive piles of MadLab leaflets to hand out."

Credit: Bob Clough

▲ *There are robots on the walls and on the tables in HAC:Man's workshop at the Manchester Digital Laboratory.*

Credit Paul Plowman

▲ *A HAC:Man participant digs around in the guts of a television.*

Credit Paul Plowman

▲ *Like most hackerspaces, laptops and soldering irons are everywhere.*

Credit: Bob Clough

▲ *Participants tinker with a MAME cabinet (see project #17).*

THE PROJECT

Participants: Bob Clough, Dave Mee, and Hwa Young Jung

HAC:Man's host organization, MadLab, had been asked to contribute to Play Space, a digital arts event with the theme of play. "During the brainstorming, the idea of making a giant Etch-a-Sketch came up," HAC:Man Clough recalled. "And since we managed to think of a pun-name almost instantly, we decided it had to be done!" Their version of the iconic game, true to its name, would have a projected display.

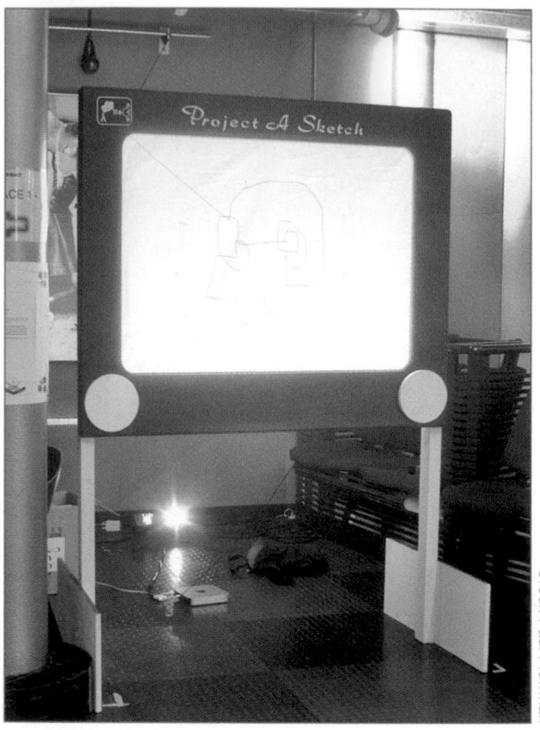

Credit Paul Plowman

▲ *The Project-a-Sketch in its home at the Manchester Digital Laboratory.*

"Step one was to build the frame," Cough wrote in a blog post. "This was done on the Shopbot CNC router at FabLab Manchester." The FabLab is sort of a hackerspace without hackers—a shop full of high-end fabrication tools available for anyone to use. (See the sidebar for more information.) Clough milled the frame out of 20mm MDF (medium density fiberboard) using a design file that Hwa Young Jung had created; he then contoured the edge to make it resemble the molded plastic curve of the original game. "We sanded the edges, painted it red, and attached some legs back at our home base, MadLab," Clough said.

Next, he built the knobs, two circles of MDF milled in the CNC and cut with a channel to allow a coach bolt to be held in place. "Acrylic washers and skate bearings were used to make sure the knobs can spin freely in the holes cut for them," Clough described.

Credit: John Honnibal

▲ *The Project-a-Sketch during startup. Because the display is projected from the rear, everything is reversed.*

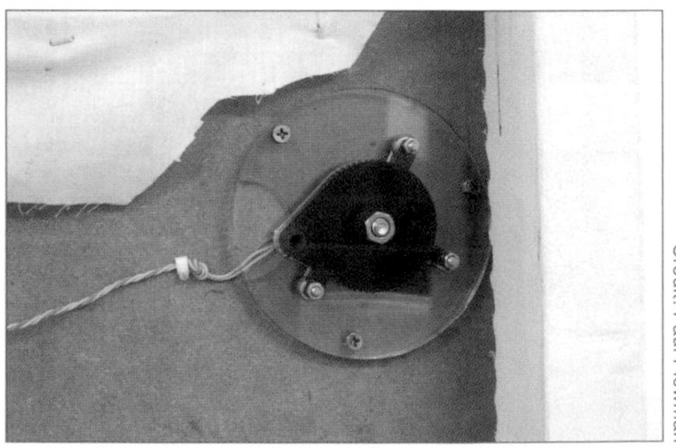

Credit: Paul Plowman

▲ *One of the project's two hand-built rotary encoders, which translates the turns of the knob into data.*

Credit: Sam Easterby-Smith

▲ *The team even laser-etched a convincing variant of the classic toy's logo.*

Credit: Paul Plowman

▲ *The accelerometer, which informs the Arduino when the frame moves. This is interpreted as a command to erase the sketch.*

Credit: Paul Plowman

▲ *The brains of the frame, the Arduino manages the data from the rotary encoders and the accelerometer.*

The knobs connect to two rotary encoders, devices that can determine how far the knobs are turned and send data accordingly. "The quadrature encoders used were low resolution, at only 16 notches per rotation, so I returned to FabLab to build a gearbox," Clough wrote in the blog post. "Using the laser cutter, I cut a 16-tooth gear to fit on the quadrature encoder and a second, 64-tooth one to attach to the knob itself." They also laser-cut an acrylic frame to hold the gears in place. "The encoder outputs 64 state changes, a far more acceptable resolution," Clough said.

Next the team added the Arduino, which interprets the data from the encoders and sends the data over a serial link to the controlling PC. They also attached an accelerometer to the frame, which the Arduino checks every 50 milliseconds. This simulates the classic shake used to erase an Etch-a-Sketch drawing. Of course, with a gigantic Etch-a-Sketch on a wooden stand, this essentially means wobbling the whole assembly back and forth.

While the body was coming together, Dave Mee wrote a Processing sketch that simulates the drawing action of an Etch-a-Sketch—a single point that draws a line on a white background. When rear-projected onto the white sheet that makes up the Project-a-Sketch's screen, the effect resembles that of the original toy.

Credit: Ben Bashford

▲ *A shot from the rear of the Project-a-Sketch shows the rotary encoders and Arduino. The line trailing down supplies power to the Arduino.*

The Project-a-Sketch was a success, engaging Play Space attendees with its fun and accessible interface. Even better, it has become HAC:Man's signature project, a toy everyone plays with when they visit the space.

Credit: Paul Plowman

▲ *The screen of the Project-a-Sketch is a simple white bed sheet.*

Credit: Hwa Young Jung / MadLab

▲ *A visitor finds the Project-a-Sketch irresistible.*

HACKERS WITHOUT SPACES

While most of the hacker organizations found in *Hack This* rent their own facilities, it's not unknown for organizations to receive donated space. There are both advantages and disadvantages to this arrangement. For instance, having a free workshop means that HAC:Man need not solicit dues from its members.

This creates a more relaxed and inclusive membership structure where keeping lists and accounts simply are not needed. People who are interested just show up, and when they stop being interested, they leave. On the other hand, a self-sufficient organization need not conform to another group's rules and whims. For instance, if MadLab decided to evict HAC:Man, they would be gone; a group renting its own space cannot so easily be evicted.

FABLAB MANCHESTER

FabLab Manchester is only one of many such facilities in the FabLab program, a creation of MIT's Center for Bits and Atoms. The program, in essence, provides high-end tools like CNC routers, laser cutters, and 3D printers to those who would not ordinarily be able to afford them.

"FabLabs have spread from inner-city Boston to rural India, from South Africa to the North of Norway," the project's site declares. "Activities range from technological empowerment to peer-to-peer project-based technical training to local problem-solving to small-scale high-tech business incubation to grass-roots research." At the latest count, there are 53 FabLabs in 17 countries.

http://fab.cba.mit.edu/about/faq/

▲ *A visitor uses FabLab Manchester's high-end 3D printer, a device no ordinary hackerspace could afford.*

▲ *Bins of tools and components await FabLab visitors. A Lego model toward the top of the picture adds a little whimsy.*

ALT.PROJECT

TWITTER DRUMMER

HAC:Man member Paul Plowman built his own project for Play Space, the conference that also featured the Project-a-Sketch. The Twitter Drummer responds to twitter messages (tweets) by banging on bongos, a cowbell, and a drum using solenoids salvaged from laser printers. The brains of the project is an Arduino with an Ethernet shield connecting it to the Internet.

See: http://wiki.hacman.org.uk/Twitter_Drummer

▲ *Paul Plowman's Twitter Drummer interprets tweets as music.*

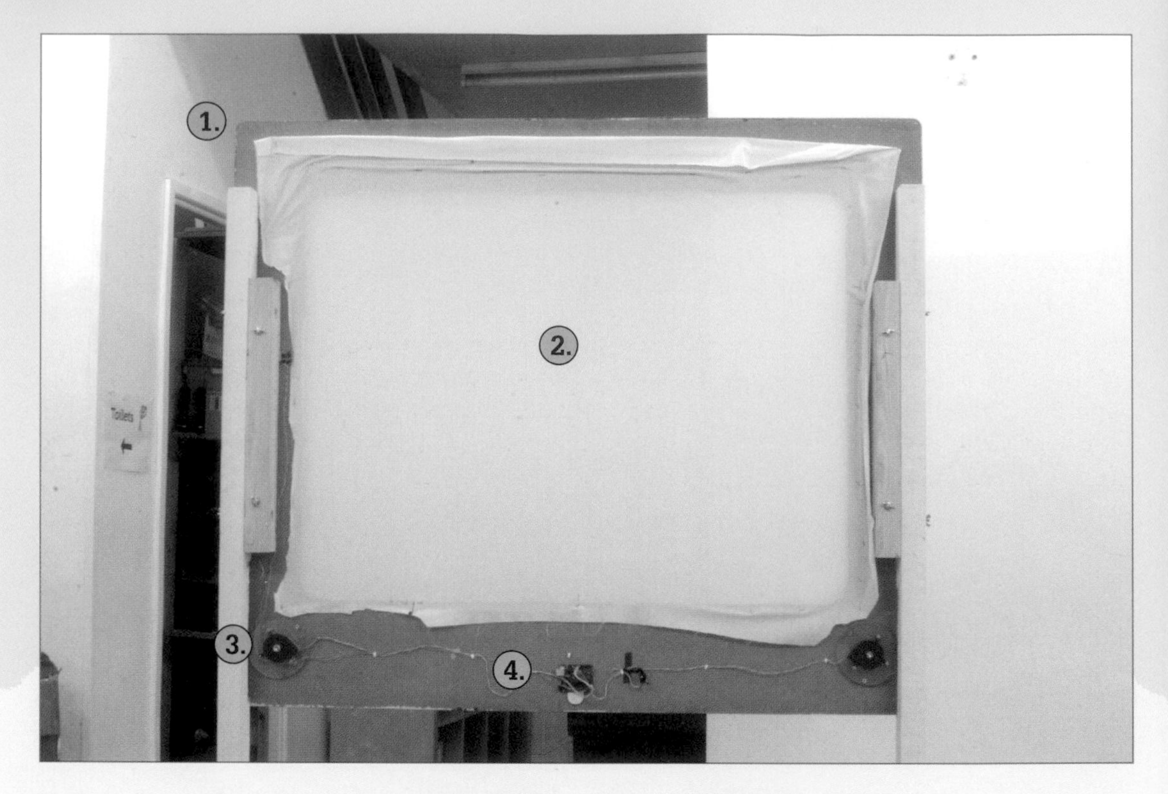

BUILD IT

The Project-a-Sketch seems simple at first glance, but it does have some complicated features. The result, however, is sure to engage both kids and adults. Here's how to build your own:

1. Mill the frame. The Project-a-Sketch team used a CNC from FabLab Manchester, but the frame could also be built using traditional wood tools.
2. Add the screen. A simple bed sheet will do.
3. Mill the knobs and attach them to rotary encoders. The project team found the 16 steps of the encoders they used to be inadequate, so they built custom gearboxes to increase this to 64 steps.
4. Install the Arduino and accelerometer. The Arduino interprets movement of the frame as a command to erase the sketch.
5. Set up the computer and projector. The computer can be any that will run Processing, while the projector must be a digital overhead projector that connects to a PC.
6. Sketch!

To see more, visit http://hacman.org.uk/project-a-sketch.

PROJECT 21

POWER RACING CAR

Milwaukee Makerspace's Grave Digger car exemplifies a trend among Midwest hackerspaces, where toy cars are retrofitted with more powerful motors capable of hauling an adult's weight and then are raced competitively.

Credit: Anne Petersen

HACKERSPACE PROFILE: MILWAUKEE MAKERSPACE

Location:

Milwaukee Makerspace

3073 S. Chase Avenue, Building 34

Milwaukee, WI 53207

milwaukeemakerspace.org

Organization type: "We are currently an LLC," president Royce Pipkins described. "We'll wait to do the 501(c)3 thing until a clear advantage arises. We don't want to commit to the educational curriculum requirements."

Founding date: December 2010

Number of members: 27

Dues: $80 regular; $40 "starving hacker"

Size of the space: 3,600 sq. ft.

Officers/Leaders:

- Royce Pipkins, president
- Tom Gralewicz, vice president/treasurer

Notable equipment: CNC router, metal working mill and lathe, drill presses, table saw, planer sanders and other woodworking tools, mig welders, tig welder, plasma cutter, oxy acetylene torch, kiln, 100Mhz 4 channel analog oscope, 34-channel USB logic analyzer, homebrew PCB etching station, large library of through-hole electronic components

SPACE DETAILS

"Our space is in an old industrial warehouse," Pipkins said. "It's kinda dirty, which makes it great to work on heavier projects. We also have a couple of air-conditioned office rooms that are great for electronics, software, and craft-type projects. The biggest thing we did was build a wall so that our neighbor tenants could get to the common bathroom without having to come through our space. The quirkiest things are the rings of bolts in the concrete floor that used to hold down cranes. There are only two or three rings, but you have to watch out for them."

Credit: Brant Holeman

▲ *Milwaukee Makerspace's welding mask logo gives a nod to the space's metalworking focus.*

Credit: Brant Holeman

▲ *A member welds a bicycle wheel.*

THE PROJECT

Participants: Royce Pipkins, Tom Gralewicz, Rich Neubaur, and Brant Holeman

You know Power Wheels, those ride-on electric cars that the cool kids drove around the neighborhood? A group of hackerspace members have begun refitting the toys with better batteries and more powerful motors, organizing races, and awarding trophies to the winners. These races, called the Power Racing Series, have been building buzz and participants since their debut in 2009.

Royce Pipkins, president of Milwaukee Makerspace, wanted to build his own Power Racing car, so he found some Power Wheels cars on Craigslist. "We got Grave Digger, Little Pink Trike, and Big Jake. I claimed Grave Digger with my presidential privilege."

Power Wheels cars are built to hold and move small children, so it's necessary to upgrade most of the internal components. "We take the tiny 1mph motors out of those things and replace them with 20mph motors," Pipkins described. "We are also fat, and as most Power Wheels will only take 130 lbs., we reinforce the plastic bodies with steel or aluminum. Lastly, we replace the tires with regular tires. We tried to use the stock tires, but found that the abuse we subject them to destroys the plastic tires with an hour or less."

Credit: Pete Prodoehl

▲ *Robotics fans tinker in one of the space's air-conditioned workrooms.*

Credit: Pete Prodoehl

▲ *Milwaukee Makerspace's big space gives members many options for large-scale projects.*

Credit: Brant Holeman

▲ *Royce Pipkins stands next to his revamped Grave Digger Power Wheels car.*

Credit: Ben Nelson

▲ *The Milwaukee Makerspace team strips down the toy. The Grave Digger's stock tires, pictured on the car, were quickly removed and replaced with sturdier rubber tires.*

The first step to building a Power Racing car, after buying the Power Wheels toy, is to weld the frame. "We had to guesstimate the proper size of the steel frame that would fit inside the plastic chassis of Grave Digger," Pipkins said. "We knew that the final fit would be a lot of cutting and fitting of plastic."

The Grave Digger chassis is a green plastic shell that comes apart along the long axis. The molded black floorboards, dash panel, hood, and side panels comprise a third piece, and the seat, rear side panels, and bed (also made of black plastic) represent a fourth piece. "I eventually gutted much of the lime green part of it to fit around the steel frame in a way that it could be taken off easily," Pipkins said. "After cutting away perhaps half of the lime green, I found I could make it work."

At some point Tom Gralewicz brought an oscillating saw to help with the shaping. "This tool was actually a big help because it cut the plastic like butter. It turned the process into fast carving rather than considered and slow cutting. It allowed me to dial the shape of the lime green plastic into near perfection by making numerous small cuts and seeing how it fit."

Before the chassis was reattached, however, Pipkins wanted to test how the new components worked. He used a milk crate for the seat and steered with a pair of vise grips clamped to the steering column. "The throttle was a winch controller that has forward and backward buttons," he said. "We used small 12V motorcycle batteries so we could just stick them someplace. The wiring on the winch controller was a little small and got

Credit: Ben Nelson

▲ *The Grave Digger's reinforced frame, with upgraded wheels and motors.*

pretty warm, but overall it served as a great testing tool."

One of the most important decisions of the conversion process was to choose the motors. "The initial choice for motors was electric jack motors. They seemed like they might be pretty torquey," Pipkins said. "However, it turned out that they had a clutch mechanism that would slip on fairly normal operating conditions for the vehicle. The clutch was just a large washer with three places where the thickness of the washer got shallow. An armature rested in those shallows and transmitted torque to the output shaft. We tried to tighten the clutch and we tried to grind out the ramps that led down into the shallow, all to make it harder for the armature to come out, but we just couldn't make it work. So we finally gave up."

Credit: Brant Holeman

▲ *Pipkins tests the strength of the frame.*

▲ *Rich Neubaur welds on the Grave Digger's motor mounts.*

▲ *The Grave Digger with the molded black plastic chassis parts being put back on after the frame is installed.*

▲ *Working late at night, Neubaur grinds on the Grave Digger's rear axle.*

Then team member Rich Neubaur brought in four old electric wheelchair motors. "They were physically much larger than the jack motors, and the gearboxes' output RPMs were a little on the low side, so we would wind up connecting the outputs to the axle via a sprocket and chain." The wheelchair motors spun too slowly, so they put a big sprocket on the motor's gearbox and a smaller sprocket on the axle. "Pretty darn inefficient, but it got me running."

After some false starts and experiments, the final configuration of the Grave Digger began to take form. Pipkins used two marine deep-cycle 12V batteries, two 30-amp 24V speed controllers, two 500W 24V scooter motors, #25 chain and sprockets, a riding lawnmower steering mechanism, and four lawnmower tires.

On race day, the Milwaukee Makerspace team headed to Maker Faire Detroit to face such hackerspace competitors as i3Detroit, CCCKC, Sector67, and Pumping Station: One. "Detroit 2010 was awesome," Pipkins said. "Working so hard and so long on something and then having it come to life and work, in front of a cheering audience, is a great experience."

Credit: Anne Petersen

▲ *Like a lot of Power Racing vehicles, the Grave Digger needed a pit stop during the race.*

Credit: Anne Petersen

▲ *On the track at Maker Faire Detroit, the Grave Digger scores a victory.*

There were breakdowns and snafus. Tom Gralewicz had tack-welded the front end of his Little Pink Trike but forgot to go back and weld it properly, so the trike ended up splitting in two during a drag race. Pipkins had his own problems, too, but ended up doing well. "I came in second overall, mostly because I wouldn't quit jockeying for position in the endurance race. My motors overheated and blew out about halfway through the race, leaving me in second. It was still a blast, though. Racing for position is a *ton* of fun. Doing it on the rattletrap you've built is that much sweeter." Gralewicz ended up taking third place overall and the team swept the Constructors' competition.

TOOL BIN

MIG WELDER

A MIG welder, found in many hackerspaces, is a tool for joining two pieces of metal. It consists of a large power supply and a nozzle that creates an arc of electricity between the surface to be welded and a consumable electrode wire dispensed through the nozzle. The heat produced by the arc melts the metal and allows the two pieces to connect.

Using a MIG welder has been likened to wielding a hot glue gun. It's considered the easiest of the welders to learn, able to work with a wide variety of alloys and thicknesses of metals. That, and the relatively inexpensive price tag ($200–$300 for a base model) make it a popular tool for entry-level welding. If you use a MIG welder, be sure to wear safety gear, including a welding mask, gloves, and protective clothing.

Credit: John Baichtal

▲ *A MIG welder is a necessity for any significant metalwork.*

POWER RACING SERIES

The Power Racing Series is a race in which makers and hackers compete with heavily modified children's electric ride-on toy cars, participating in a variety of races in the hopes of winning trophies and bragging rights.

The series is abbreviated PPPRS, a salute to the Power Wheels advertising jingle from the '90s, where an exited voice sings "Pow-Pow-Power Wheels." "We put that kind of enthusiasm in our acronym: PPPRS—Pow-Pow-Power Wheels Racing Series," series organizer Jim Burke said. "Of course, for legal reasons, we don't actually say 'Power Wheels Racing,' but our abbreviation is our nod to that ad."

Teams can spend up to $500 to customize a stock Power Wheels, Little Tikes, or similar powered child's toy but add bigger batteries, faster motors, and other parts. They face off in drag races, road courses, and a climatic 75-minute endurance race to claim the championship.

The series is deeply linked to the Midwest U.S. hackerspace scene. "Way back in 2009, we were having a good time at PS:One when a Power Wheels rescued from a dumpster made its way into our newfound space," Burke recalled. "The instant nostalgia of the plastic pink Jeep made us giddy with the idea of driving it around. After a few drives around I made the mistake of suggesting it would be fun to race these hapless machines."

A few months later, PS:One's race was held in an empty lot. "It was small, with just six cars entered, and only a few score of spectators showed up to what would be an amazingly entertaining August afternoon." Other spaces participating included i3Detroit, Sector67, OmniCorp Detroit, CCCKC, HackPGH, and All Hands Active.

Credit: Brant Holeman

▲ *Royce Pipkins's Grave Digger (left), with Pink Trike and Big Jake. To the right is Pumping Station: One's Micro CHIP police bike.*

@mkemakerspace
Milwaukee Makerspace

@I3Detroit GraveDigger's gonna dig ur grave Big Jake's gonna push u in n cover u up then Little Pink Trike gonna peel out on top of ur grave

10 Jun via TweetDeck

▲ *Milwaukee Makerspace's smack talk on Twitter mentions three of the group's Power Wheels vehicles.*

The series has grown since its inception, with races featured at Maker Faire Detroit 2010 and the Kansas City Maker Faire the following year, and the number of participating cars has increased from 6 to over 20. "We're getting so many requests now that we will have to cap the field number next year and limit how many entries a hackerspace can enter," Burke said.

For more information, check out powerracingseries.org.

BUILD IT

Building a Power Racing Series car can be an expensive proposition, with teams investing up to $500 (the rules' limit) per car. Here's how to make your own:

1. Buy a Power Wheels car. You can often find them at yard sales and on Craigslist, sold by parents who don't want the hassle of replacing the cars' batteries.
2. Remove the plastic chassis.
3. Weld a new frame. Unless you weigh 80 lbs., you'll want a stronger frame to hold the diver.
4. Connect new motors. The stock motors top out at 1mph, so you'll need something with a little more zip if you want to compete.
5. Install batteries. The Grave Digger packs twin 12V deep-cycle marine batteries, but the motors you use determines what sort of batteries you will have to buy. Pipkins put his batteries under the Grave Digger's hood.
6. Add tires. Power Wheels stock tires are cheap plastic shells, so you'll need more robust wheels like those found on mobility scooters or riding lawnmowers.
7. Reuse or replace the toy car's steering mechanism.
8. Reattach the plastic chassis.

PROJECT 22

PARTY LAND PINBALL GAME

Felipe Sanches always loved playing pinball on the computer. Eventually, he decided to build his own pinball table, one that looked and acted just like the one on the screen.

HACKERSPACE PROFILE: GAROA HACKER CLUBE

Location:

Garoa Hacker Clube

Rua Vitorino Carmilo, 459—Santa Cecília

São Paulo, SP, Brazil
CEP 01153-000

http://garoa.net.br/wiki/Garoa_Hacker_Clube:About

Organization type: Nonprofit

Founding date: July 2010

Number of members: 20

Dues: R$40/month

Size of the space: 12m^2

Officers/Leaders:

- Felipe "Juca" Sanches, president
- Gustavo "Aylons" Bruno, hardware director
- Mauro Baraldi, software director
- Hugo "Agaelebe" Borges, treasurer
- Rodrigo "Pitanga" Rodrigues da Silva, secretary-general

Notable equipment: MakerBot, drill, soldering irons, multimeters, oscilloscope, Dremel, power supplies, signal generator, and a hot air gun

SPACE DETAILS

Garoa Hacker Clube is located in a castle-like building known as the Casa da Cultura Digital, along with several other technology-related organizations. Sanches has been customizing the walls of the space with stenciled images of electronic components, logos, and other images (see Alt.Project, below.)

Credit: PERMISSION TK

▲ *Garoa Hacker Clube is located in a beautiful, castle-like building that houses several organizations.*

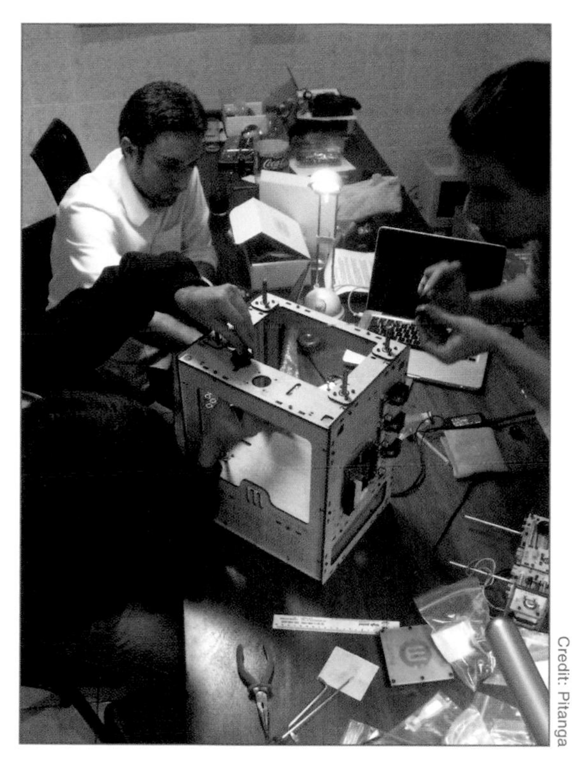

Credit: Pitanga

▲ *Members assemble the space's MakerBot.*

Credit: Aylons Hazzud

▲ *Garoa Hacker Clube's electronics bench awaits the next project.*

Credit: Felipe Sanches

▲ *GHC members have customized their space with stickers and stencils.*

Credit: Fabricio C. Zuardi

▲ *The half-finished table with printed paper graphics, awaiting the transparent acrylic pieces that will cover them.*

THE PROJECT

Ever since he was 12, playing a pinball simulator on the PC, Felipe Sanches dreamed of creating his own pinball game. At first, Sanches focused more on the software end. "I wanted to know how computer programs are made, and I did not have Internet access so I had to learn by myself by tinkering with whatever was available in a typical 386 computer."

As a high school student, Sanches began the formidable task of reverse-engineering the simulator, called Pinball Fantasies, learning assembly language along the way.

▲ *Sanches's game board shows the pencil marks of planned improvements.*

▲ *Home-built elements mix with actuators salvaged from broken pinball games.*

Pinball Fantasies was a game program released in 1992; it came with four virtual pinball tables that could be played just like regular pinball games, with algorithms determining where the ball went. Sanches's original goal was to create a nonproprietary version of the game in which anyone could develop new pinball layouts using the game engine.

In college, however, Sanches stopped working on the project. "I had an idea of something much more interesting," he said. "I had gathered enough technical info about the game—so much knowledge about the internals of it—that I figured out it would be possible to hack it into a controller for a real machine."

▲ *Sanches's controller board allows an Arduino to control several components including lamps and LEDs.*

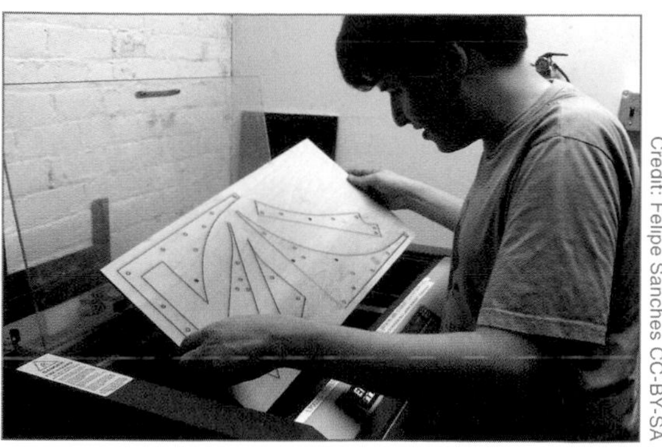

▲ *Plywood elements come out of the laser cutter at New York's NYC Resistor.*

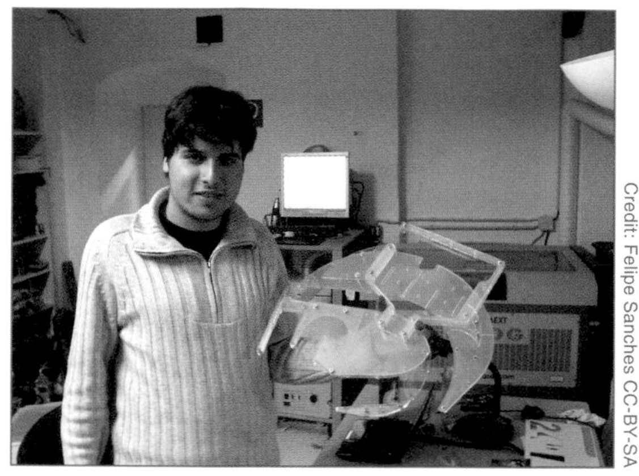

▲ *Sanches shows off some laser-cut acrylic pieces he made at Metalab in Vienna, Austria.*

Around 2006, he began building a physical version of Pinball Fantasy's Party Land game. "I was trying to do something that I had never done before," he wrote in a blog post. "Building a homebrew pinball machine sounded to me like a very ambitious project."

It was also a solitary challenge, or so it seemed to him at the time. "I hadn't even heard of other people who had done similar projects," he said. "I felt alone and obliged to come up with my own solutions to overcome each of the challenges involved in the execution of the project." Since then, however, he has made several contacts with other hackers working on pinball games.

▲ *The tangle of wires connecting the various actuators to the Arduino underscores the complexity of the project.*

Sanches's system involves a Pinball Fantasies game on a PC running the DOS emulator DOSBox. "I modified the DOSBox emulator so that it would detect some specific patterns in the Pinball Fantasies runtime behavior and, based on that, infer what's going on in the game and use that info to make the real machine behave accordingly," he explained. The computer is connected to an Arduino via USB, sending serial data to the microcontroller to manage the lights and actuators of the physical game. He designed his own lighting controller board using gEDA, a suite of free PCB layout utilities.

In addition, Sanches built a LED matrix to display game-authentic animations and messages sent by the PC. Music and sound effects, on the other hand, come straight from the PC's speakers.

During the project's lengthy build process, Sanches had the opportunity to travel both to Europe and the United States and visit a number of hackerspaces, including NYC Resistor, Noisebridge, Metalab, and Brmblab. He gave presentations on pinball hacking and used these spaces' laser cutters to make acrylic pieces for his project. "It was mostly structural parts for the game," he said, "the kind of thing that is not available in old machines because it is highly dependent on the layout of the specific game."

Along the way, he met several pinball enthusiasts and visited a pinball museum, acquiring a number of difficult-to-recreate pinball actuators like pop-ups and bumpers, harvested from old machines.

Even after all the work Sanches has put into the project, the game is not quite playable.

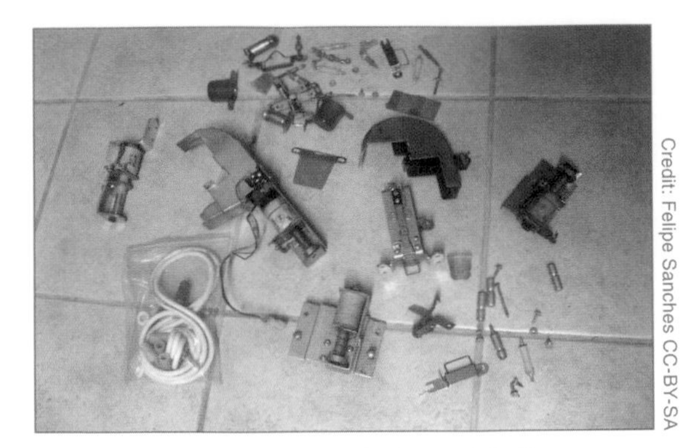

▲ *Parts harvested from a broken pinball game saved Sanches from having to develop his own.*

▲ *The acrylic pieces.*

The main thing he has to complete is the system for managing the solenoid bumpers, pop-ups, and other game effects. "I need to drive something around 14 solenoids with a reduced subset of I/O pins of the Arduino," he said. "The rest of the I/O pins are already used for controlling lamps, switches, and the dot-matrix-display." Eventually, Sanches hopes to develop a universal pinball controller board so other pinball fans can create their own games.

ALT.PROJECT

HACKERSPACE STENCILS

Sanches has decorated Garoa Hacker Clube with several stencils of hacker-related topics, including logos and pictures of electronic components. He created these by placing a sheet of exposed x-ray film over a computer monitor displaying the image to be stenciled. The design is traced with a marker and then cut out with a razor blade, leaving bridges that connect every part. Then the mask is taped to the wall with newspaper surrounding it to catch the splatters. All that remains is to spray the mask with paint.

Credit: Felipe Sanches

▲ *One of Sanches's stenciled images, showing a transistor.*

Credit: Felipe Sanches

▲ *Sanches created his stencils by tracing images onto old x-ray film and then cutting them out.*

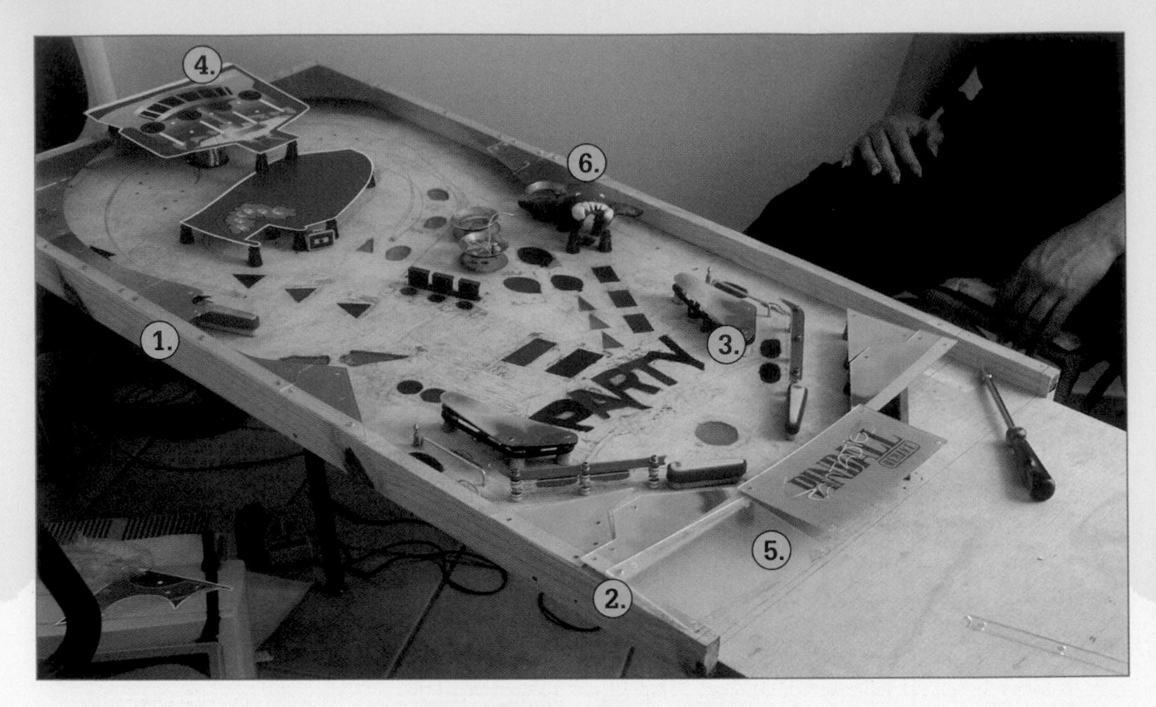

BUILD IT

To build a pinball game, you'll need a lot of separate elements all working together ranging from solenoids to LEDs to flippers:

1. Design the game's playfield. Sanches elected to re-create the PartyLand game from Pinball Fantasies.
2. Build the table. In its most basic form, it's a plywood box with the gamefield cut into it.
3. Add gates, flippers, pop-ups, and other actuators. Sanches reused parts from old pinball games because of the difficulty of re-creating all of the elements from scratch.
4. Design and print decorations. You can do this on the computer, design the shapes in a design program like Inkscape, and then print on a color printer.
5. Build ramps and other acrylic elements. Ideally, these should be laser-cut. Sanches provides the Scaled Vector Graphic (.svg) files on his Google Code site, listed in the following section.
6. Install bonus extras, like the monster head Sanches formed out of modeling clay.
7. Load DOSBox and PartyLand software on your computer. (Not shown.)

FURTHER READING

Project Site:

http://code.google.com/p/how-to-build-a-pinball/

Felipe Sanches' website:

http://jucablues.blogspot.com/

PROJECT 23

STORE FRONT MUSIC

Responding to a challenge, the members of Los Angeles' CRASH Space transformed the front of their building into an interactive musical instrument.

HACKERSPACE PROFILE: COLLABORATIVE RESEARCH ASSOCIATION OF SOCIAL HACKTIVITY

Location:

CRASH Space

10526 Venice Boulevard

Culver City, CA 90232

www.crashspace.org

Organizational type: California corporation working on becoming a 501(c)3 nonprofit

Founding date: December 2009

Number of members: 31

Dues: $108/month for "Super Users" or $37/month basic membership

Size of the space: 1,400 sq. ft.

Officers/Leaders:

- Sean Bonner, president/CEO
- Carlyn "carlynorama" Maw, vice president/COO
- Tod "todbot" Kurt, chief geek/CTO
- Mike "mikeout" Outmesguine, treasurer/CFO
- Justin "outlawpoet" Corwin, secretary

"We do a lot by member consensus," board member Carlyn Maw described. "There is the board, of course, but any given member can choose to take on as much responsibility as they can handle. Collaborative Research Association of Social Hacktivity (CRASH Space) is meant to be a platform for the members to create what they want out of it. The board builds the sandbox, the members make the sand castles."

NOTABLE EQUIPMENT

Laser cutter, four-axis CNC mill, a bench lathe, three drill presses ("trying to get down to two"), two HP network analyzers, "a colony of MakerBots," oscilloscopes, a bandsaw, a two-ton press, a high-temperature furnace, a kiln, an arc welder, a CAD workstation, and a large-format color printer.

SPACE DETAILS

"The building that CRASH Space lives in is a combination of a house and an office and a machine shop," Maw said. "It's been a lot of things and we're just the latest. The facade has a nice 1970s feel to it with large picture windows where we can display what we've been working on. It is on a prime street with lots of yummy food in walking distance. The landlord liked us because we were prepared to take it as-is without a lot of high maintenance requests for carpet upgrades, etc. It has multiple rooms so folks can get machine shop projects done, while someone works in the office, while another group has a class...."

Credit: Theron Trowbridge

▲ *CRASH Space's facility has large windows that the organization uses to show off their work. In front of the windows you can see the PVC framework that houses the ultrasonic sensors.*

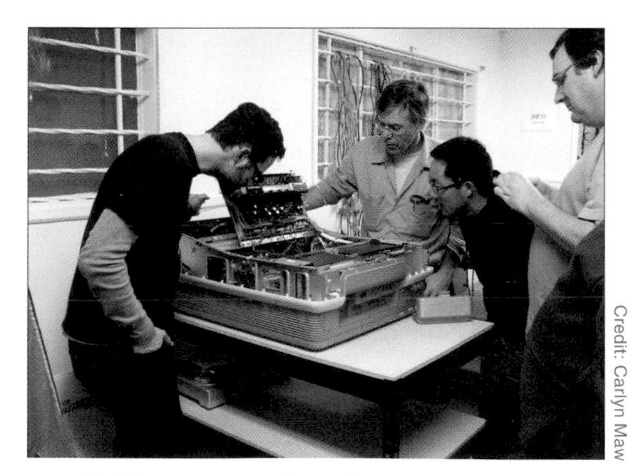

Credit: Carlyn Maw

▲ *CRASH Space members dig into an obsolete projector. The organization reserves "Take-Apart Tuesday" for disassembling junk and learning how it was put together.*

Credit: Theron Trowbridge

▲ *A lecturer gives a presentation on "The Joy of Wire"—a topic most hackerspaces would love to learn more about.*

THE PROJECT

Credit: Daryll Strauss

▲ *The musical instruments of Store Front Music can be seen through CRASH Space's front window.*

In early 2010, members of CRASH Space in Los Angeles were contacted by a production team working for video site Vimby, which wanted to feature the hackerspace in a program.

"CRASH Space has been approached by a number of production companies in a 'wouldn't you like to be on TV' sort of way," Maw recalled. "It has something to do about being in LA."

But this one was different. For one thing, carmaker Scion was prepared to contribute $3,000 to help CRASH Space create a project for the show. The fact that legendary hackerspace evangelist Mitch Altman had signed on as host only reinforced the legitimacy of *Take On the Machine*, as the program was called.

Credit: Carlyn Maw

▲ *CRASH Space members gather around a workbench to tinker with the project's electronics.*

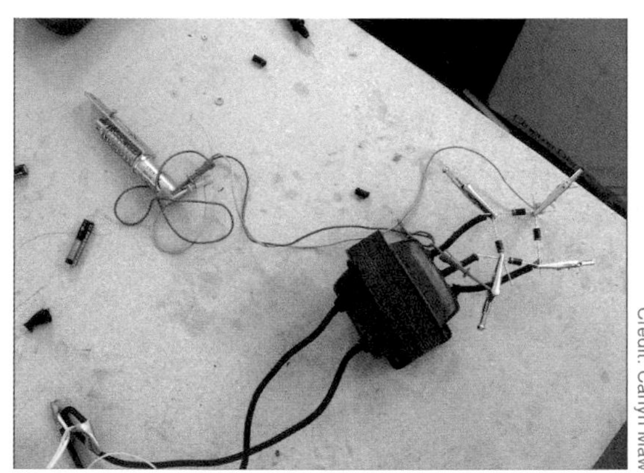
Credit: Carlyn Maw

▲ *The critical element of the project, the solenoid (upper left), gets hooked up to a transformer that powers it.*

Credit: Carlyn Maw

▲ *Perfboarded modules await assembly into the finished project.*

A meeting was called, and the members agreed to participate. The next step was to decide on a project.

"It slowly narrowed down to what we call Store Front Music," Maw said. "Store Front Music is an installation piece that turns the front of our hackerspace into a musical instrument."

CRASH Space rents out a building with a huge front window, so they proposed to create microcontroller-connected percussion instruments that would trigger whenever someone walked past the window.

The core of the project is a bank of Maxbotix LV-EZ4 ultrasonic sensors housed in a PVC frame jutting from the building's window planters. The sensors trigger eight SOL-58 solenoids, which use lengths of plastic rod as drumsticks —the team called them "thwackers."

"The instruments were a lot of fun to make, and we had very theatrical thwackers," Maw said.

For noisemakers they use bottles, terra cotta pots, heat sinks, and an old guitar. The resulting sounds are picked up on microphones and played on outdoor speakers. Meanwhile, "UFO-inspired lighting" shines through the glass.

"One of the reasons this idea won out is because it is very modular," Maw said.

Individuals could work on different parts of the project: one person could tackle the window installation, another member could focus on lighting, and everyone worked on the musical instruments.

"Music is highly accessible," Maw said. "Everyone wants to make noise—well except ninjas, but even they think about it."

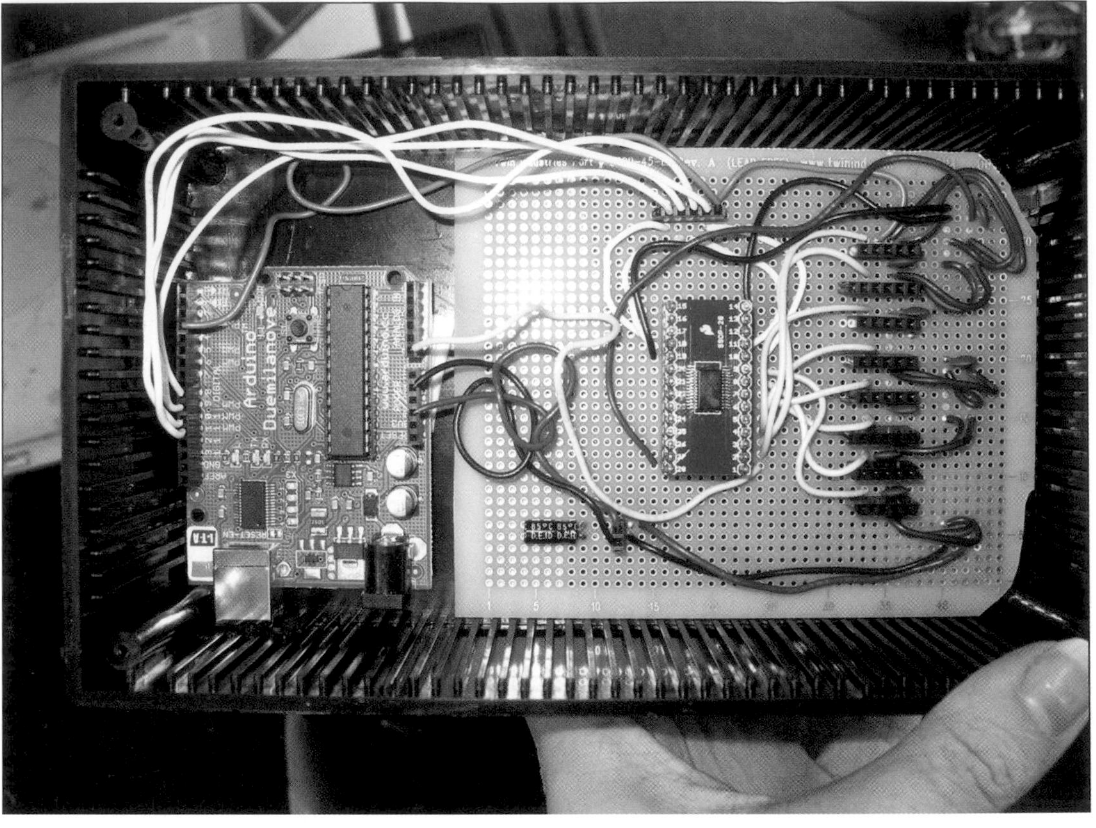

Credit: Carlyn Maw

▲ *This weatherproof project box holds the Arduino and a multiplexer that handles eight simultaneous sensor feeds.*

Writing software was a big endeavor. With so many sensors and actuators needing to cooperate, the team employed multiplexers—boards that accept input from many devices and combine their signals into one feed.

"The first step was to write code that could read eight analog ins from the multiplexer and spit them out serially," Maw said. "We did the original testing with potentiometers and other analog sensors that we knew worked."

One of the *Take On the Machine* rules was that the project had to reference a commercially released motion picture. CRASH Space chose *Close Encounters of the Third Kind*, based on the movie's musical references. They even programmed in an Easter egg—if you trigger all eight sensors within a two-second span of time, the speakers blare the classic five-note *Close Encounters* theme. A Mac Mini and multiple Arduinos coordinate everything.

Power turned out to be a stumbling block. There were eight solenoids, and each one drew two to three amps when fired.

Credit: Carlyn Maw

▲ *One of the ultrasonic sensors receives a last-minute inspection before it is installed in its PVC housing.*

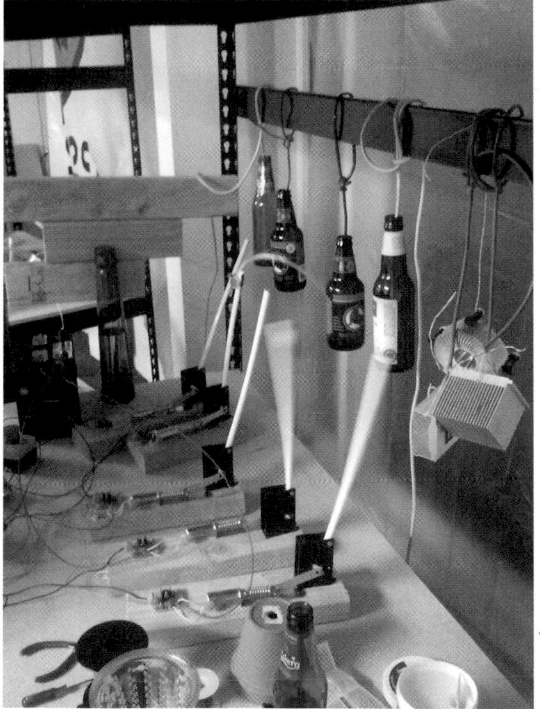

Credit: Daryll Strauss

▲ *The solenoid-driven thwackers strike bottles and heat sinks in response to sensor stimuli.*

"And they were all constantly firing," Maw said. "We ended up using a lawn mower battery with a car battery charger hooked up to it the whole time." The battery provides DC power that drives the project, while the car charger uses wall current to keep the battery charged.

As the project neared completion, the team faced two deadlines:

- The first was the conclusion of the *Take On the Machine* challenge, which they easily beat.
- The second deadline was a self-imposed one—they were going to throw a party to show off their creation to the public.

"A lot of people think of hackers and hacking as antisocial. But curiosity is contagious," Maw said. "For me, a large part of hacking is the tremendous joy in figuring out how things tick, finding new uses for common stuff, and sharing that information. The more people involved, the more knowledge, the more insights. It's ultimately a very social thing, and I think this project reflected that."

With the contest over, members hope to keep Store Front Music operational, with plans to add ever more complex components such as compressor-driven wind instruments, Wii remotes, and a laser harp.

TAKE ON THE MACHINE

To members of the scene, hackerspaces' rapid rise in popularity seems to have struck with the suddenness of a lightning bolt, but the phenomenon is only slowly permeating the public consciousness. So, when video site Vimby and carmaker Scion announced a series of web videos featuring a challenge between five hackerspaces, it generated a lot of excitement. Was this a sign that the movement had gone mainstream?

The show *Take On the Machine* consisted of a competition between five hackerspaces. Each organization received $3,000 and three weeks to create a cool project. Each had to fulfill three requirements:

- The project must repurpose a consumer product.
- The project must be interesting and fun to the average person.
- The project must reference a commercially released movie.

The hackerspaces participating included

- CRASH Space, Los Angeles, CA
- NYC Resistor, New York, NY—featured in Project 2: Sudo Make Me a Sandwich Robot
- Artisan's Asylum, Boston, MA
- theTransistor, Provo, UT—featured in Project 17: MAME Cabinet
- Pumping Station: One, Chicago, IL—featured in Project 11: TARDIS Photobooth

Giving the show a charismatic host as well as a figure well known in the hackerspace scene, Vimby recruited legendary inventor and electronics guru Mitch Altman to act as judge. Altman met Vimby producers at a chance encounter at Philadelphia's Hive76 hackerspace. A few weeks later, the producers contacted Altman via Twitter and asked him to join the project.

"I really enjoyed hosting the series," Altman said. "I'd never done anything like it before. This was basically a short 'reality' TV show."

The production crew worked with Altman to write a script, but it was all stuff he would have said anyway.

"Having things come out friendly and spontaneous when it is actually scripted was way interesting for me," Altman said. "Turns out I like doing this!"

Ultimately, Altman chose NYC Resistor's project, a cocktail-dispensing slot machine, as the winning entry.

To learn more about *Take On the Machine*, go to vimby.com and search on the show's name.

TOOL BIN

PULL-TYPE SOLENOID

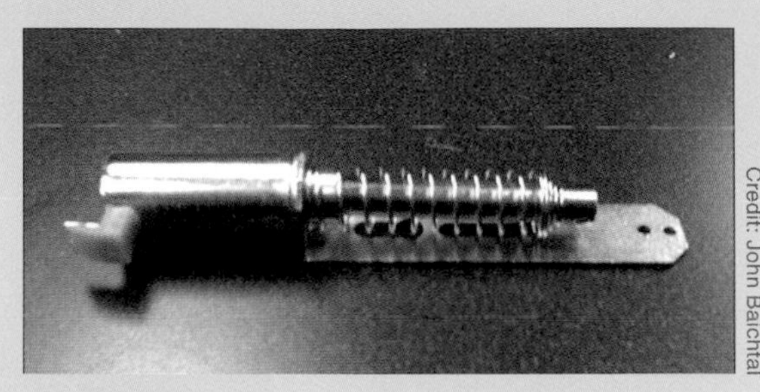

Credit: John Baichtal

▲ *The SOL-58 solenoid translates electricity into linear motion.*

Perhaps the most critical component of Store Front Music are the devices that swing the strikers—these are called *solenoids*. They're like motors, but instead of spinning, solenoids either push or pull a rod in response to an electrical signal. The SOL-58 draws 12 volts at two or three amps, and its 10% duty cycle means that it can be used only intermittently; potentially damaging heat can build up if it's triggered too often. For the next version of Store Front Music, the CRASH Space team is thinking of switching to pneumatic strikers.

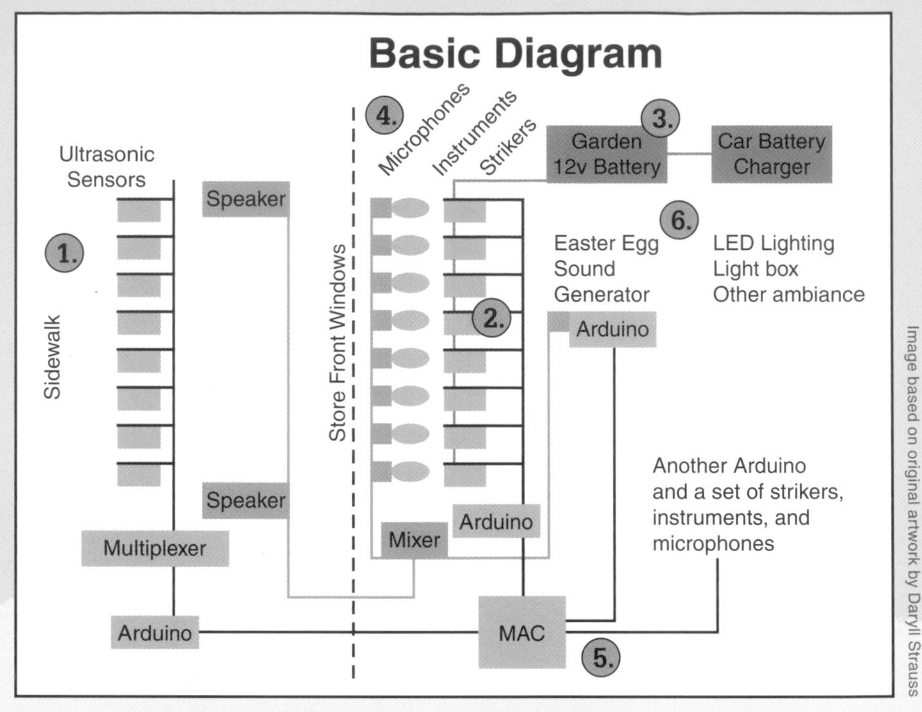

▲ *Eight sensors trigger eight strikers, and the resulting tunes are picked up by eight microphones.*

BUILD IT

The complexity of Store Front Music, with dozens of components ranging from solenoid strikers to ultrasonic sensors, makes the project hard to visualize. The following diagram illustrates how CRASH Space organized the project.

Building a sensor-driven, interactive musical instrument can be challenging, but following these steps can take out some of the guesswork.

1. Wire up the sensors and the Arduino that controls them. If you have several sensors, you might want to look into a multiplexer. The CRASH Space team used a BOB-09056 multiplexer from Sparkfun. This allows you to connect up to 16 analog sensors to a single microcontroller. Also, if this assembly is to be located outside, be sure to protect it from the elements.

2. Build the solenoid assemblies. Each consists of a solenoid, a laser-cut plastic enclosure, and a striker cut from a plastic coat hanger. An Arduino coordinates the striker action.

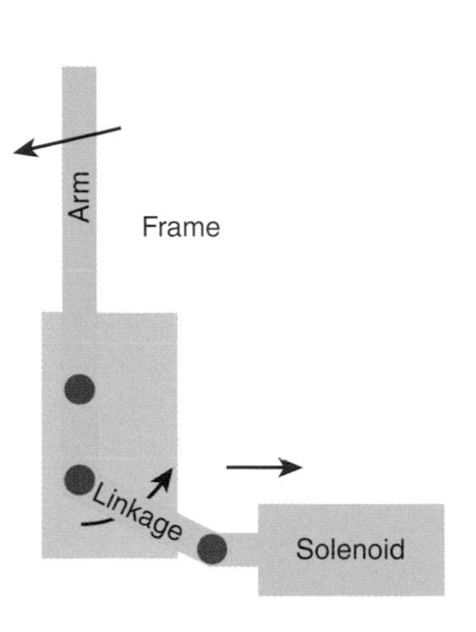

All the connections were able to pivot that allowed the arm to swing in an arc with the linear motion of the solenoid.

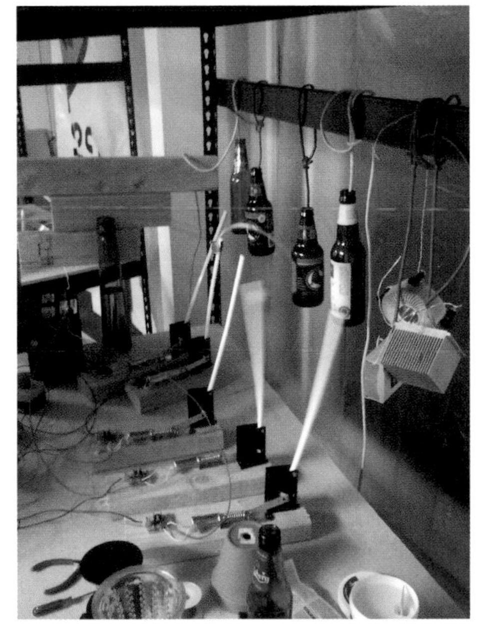

Image based on original artwork by Daryll Strauss

▲ *The pivoting connections allowed the arm to swing in an arc.*

3. Choose a power source. The solenoids draw too much current to get power from the Arduino, so they'll need a separate power source. CRASH Space used a 12V lawn-mower battery plugged into a car charger. This assembly turns the AC of wall current into DC that the components use.
4. Set up microphones to pick up the sound of the instruments and transmit them outside. CRASH Space used a custom sound mixer to combine the separate mic signals into one feed.
5. Attach the computer. CRASH Space used a Mac Mini, but you can use any one that interfaces with Arduinos.
6. Set up the Easter egg sound generator. This is a separate Arduino connected to the Mac Mini.

FURTHER READING

You can find the project code here:

http://code.google.com/p/crashspace-sfm/source/browse/trunk/docs/

PROJECT 24

WHEELCHAIR ROBOT

Hackers at San Francisco's Noisebridge hackerspace turned an electric wheelchair into a laptop-controlled robot.

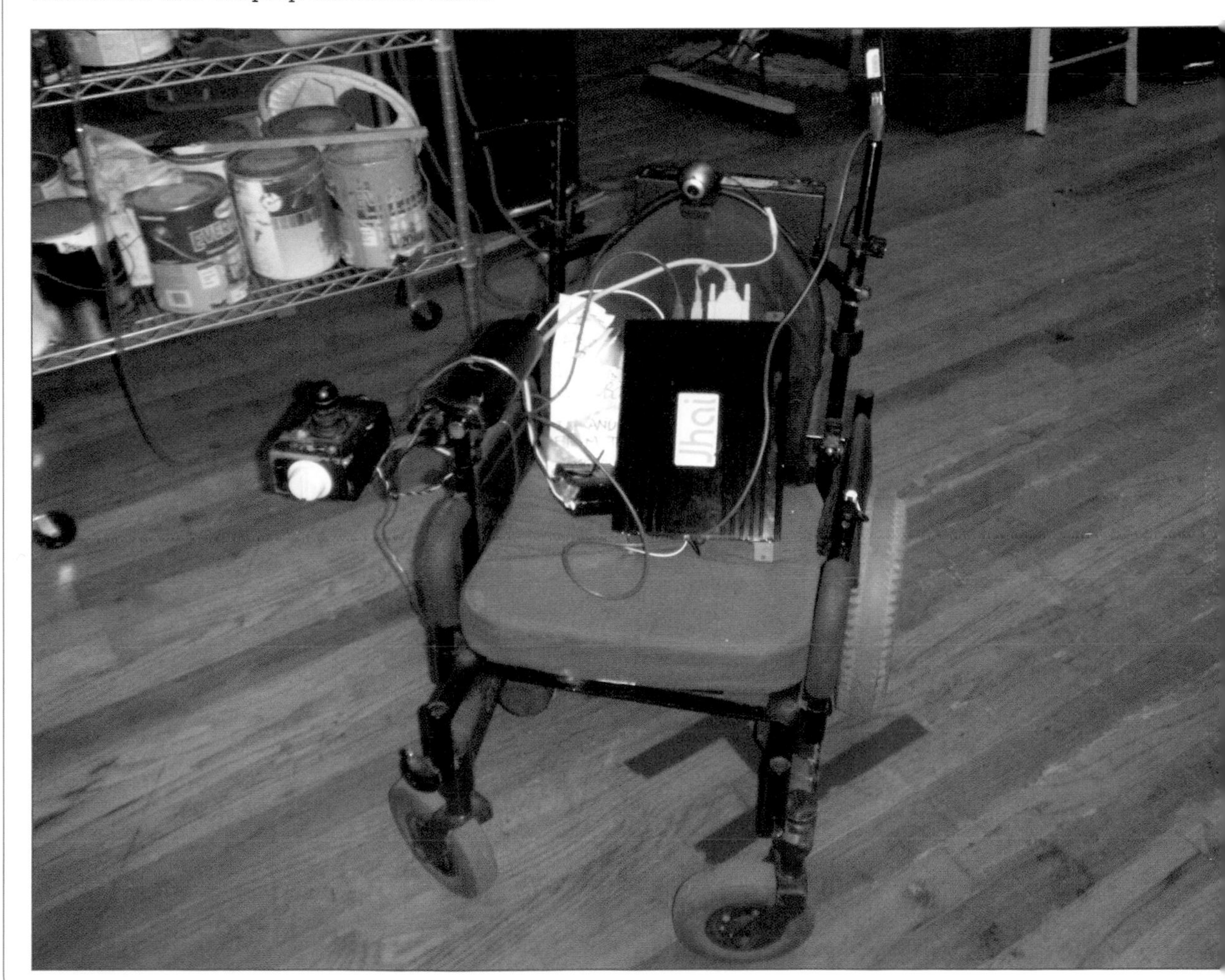

HACKERSPACE PROFILE: NOISEBRIDGE

Location:

Noisebridge
2169 Mission Street
San Francisco, CA 94110
noisebridge.net

Organizational type: 501(c)(3) nonprofit

Founding date: October 2008

Number of members: 48

Dues: $80 regular, $40 "starving hacker"

Size of the space: 5,600 square feet

Officers/Leaders:

- Jonathan Lassoff, president
- Kelly Buchanan, CFO
- Seth Schoen, secretary
- Miloh Alexander, board member
- Rachel Lyra Hospodar, board member
- Danny O'Brien, board member
- Jonathan Lassoff, board member
- Al Sweigart, board member

Notable equipment: Laser cutter, fully equipped kitchen, two classrooms, fully equipped darkroom, library, wood and metal shops, CNC mill, electronics area, industrial sewing machines, optical and scanning electron microscopes, electroencephalograph machine

SPACE DETAILS

"We have a big, bright, sunny space in the heart of the San Francisco Mission District with a great atmosphere for working and socializing alike," Noisebridge member Lilia Markham described. "The space is open 24 hours a day, open to absolutely anyone and everyone who can ring the door buzzer. There are also several methods for hacking your way into the physical space. Occasionally, that means interesting things happen during late hours when people are hitting their Ballmer curve and being really productive. The majority of the space has a large open floor plan, which encourages cross-pollination of technical and cultural ideas. I believe our most valuable resource is the workspace that is distributed throughout the space. Tables and desks and rolly chairs make things happen."

Credit: Paul Szymkowiak

▲ *The main entrance of Noisebridge, in the heart of San Francisco's Mission District.*

Credit: Paul Szymkowiak

▲ *Most of Noisebridge is an open space where lots of different activities take place.*

Credit: Fumi Yamazaki

▲ *Two Noisebridge members work on a project.*

Credit: Fumi Yamazaki

▲ *Noisebridge's huge library; above it is an LED matrix playing Conway's Game of Life (see sidebar).*

THE PROJECT

Participants: Jake Sternberg, Lilia Markham, Paul Klastner, Rachel McConnell

Like many hackerspaces, San Francisco's Noisebridge receives occasional donations of electronic junk that the donor figures SOMEONE will want to hack on. Often the junk merely gathers dust; occasionally a person or group will do something cool with it. A group of Noisebridge members took a donated electric wheelchair and turned it into a robot.

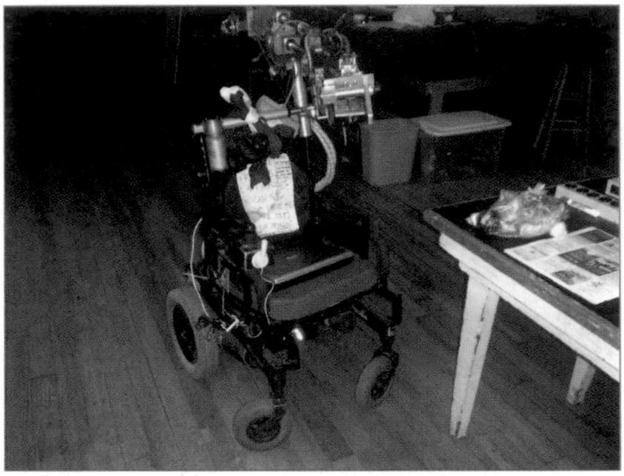

Credit: Jake Sternberg

▲ *The chair, dressed up as Santa for Halloween, sits in Noisebridge. The sign urges potential tinkerers to read the wiki before diving in. The giant robotic arm was eventually removed.*

Credit: Jake Sternberg

▲ *Noisebridge visitor Steve Castellotti adds EEG capability to the robot.*

"My friend John has a day job repairing electric wheelchairs and other equipment for people," Noisebridge member Jake Sternberg explained. "As a result, he has extra electric wheelchairs. He gave me one which was not wanted by anyone, and I thought it was the obvious thing to do: put a computer on it and hook it up to the joystick."

The idea being that if a person can drive the wheelchair by pushing a joystick, then a computer can send instructions to do the same thing, if hooked up to the joystick's ports. "I brought it to Noisebridge and put a sign on it so it wouldn't get disassembled or thrown out," Sternberg recalled. He is a hardware hacker with little interest in software, so he needed a partner to finish the robot. "It sat there for a long time, and I approached many people trying to get them to take interest in it."

Credit: Jake Sternberg

▲ *The wheelchair stripped down to its motor and giant battery packs. The two car batteries together weigh over 100 pounds.*

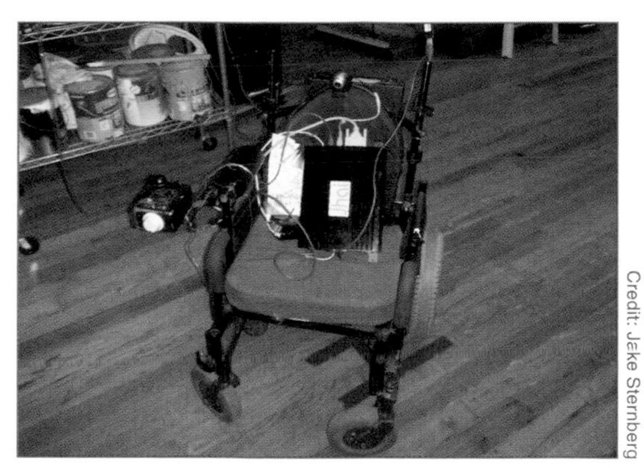

▲ *The wheelchair with a new computer, replacing its accustomed laptop.*

▲ *Team member Paul Klastner makes some last-minute tweaks before the beginning of Maker Faire.*

Sternberg finally recruited Paul Klastner to help make the robot happen. After the two built the board that interfaced between the computer and the chair, Klastner coded up a Python script that controlled the joystick. "I didn't know Paul very well at all, and I was surprised at how un-flakey he was about getting it working, so the two of us worked very hard on it to get it ready for Maker Faire 2010, which we did," said Sternberg.

The original idea was to make the wheelchair into a telepresence robot, where the operator looks through a camera and operates the robot wirelessly. The collaborators brought the chair to Maker Faire, a massive DIY convention held annually in the Bay Area.

"In theory, since there is Internet throughout the Maker Faire, the thing could scoot around the entire fairgrounds providing its drivers with entertainment safe from mortal danger," Sternberg recalled. "In reality, the wireless networking band is so heavy with traffic at Maker Faire that it's impossible to even get a connection, let alone a low-latency high-bandwidth connection to your robot wheelchair."

After the Maker Faire, Sternberg met Lilia Markham, who quickly took over the software end and loaded the wheelchair's laptop with code. "I was very impressed when I realized

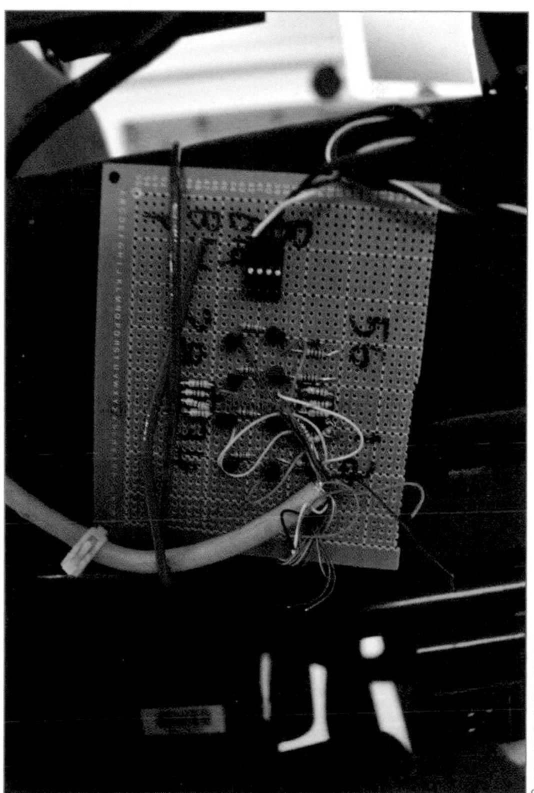

▲ *The robot's control board interfaces between the wheelchair's electronics and the computer.*

that she basically has a PhD in robot navigation and is the best coder I've ever met," Sternberg said.

Because of the project's complexity and the wheelchair's weight and power, leaving the chair unattended at the hackerspace presented some potential problems. As a complicated, heavy, and motorized hunk of metal, an injury or property damage seemed possible if someone turned on the robot but didn't know how to turn it off. "I mean, its power source is two full-sized car batteries, about 100 pounds by themselves—so the whole robot weighs more than a person," Sternberg explained. "And when set to full-speed mode, it can get going pretty fast. Most of the time when people are playing with it, there's a lot of laughter happening, and that's great, and I would hate for that to be spoiled by someone staring at their bones sticking out through their skin." Sternberg and Markham wrote a wiki page on Noisebridge's website explaining how to start up and shut down the robot.

Describing the chair presents a challenge because it has continued to evolve since the project's inception. An LCD monitor was mounted on the seatback. They installed an Arduino to manage the power, shutting down the unit if its batteries charged down too low, preventing damage to the cells. At one point Markham installed a toy rocket launcher to the chair's back while Sternberg was out of town. At various times it has also featured a gigantic robotic arm.

Credit: Jake Sternberg

▲ *The project's customized Arduino manages the chair's power in order to protect its giant batteries.*

Credit: Jake Sternberg

▲ *Another Arduino connects the wheelchair's serial connector to the computer's USB port.*

One time a visitor brought an EEG-controlled toy helicopter to Noisebridge. An EEG is a headset that interprets brain signals and can be used to control a toy…or a wheelchair. "I was there working on the robot, and it seemed obvious that we should connect the two things together and make a brain-controlled wheelchair." They quickly added EEG control to the robot and drove it around the space by brainpower.

"This project is not complete and may never be complete. I don't even know what the goal is, although having fun is definitely a primary goal. My personal goal is to facilitate software development on a real robot, and knowing that software people tend to be turned off by hardware (pun intended), I thought that by making this hardware, I could see some interesting results."

Credit: Jake Sternberg

▲ *The inside of the wheelchair's joystick shows how the Noisebridge team spliced in wires that bring instructions from the computer.*

The latest development involves the installation of a donated Kinect, an add-on for the XBox gaming system that tracks a human body and uses the data to control the game. Since its release, the Kinect was hacked for use on multiple systems and many spaces, including Noisebridge, began experimenting with the gadget.

DO-OCRACY

Many hackerspaces operate as something of a benevolent anarchy, where decisions are made by those with the most time and enthusiasm. This philosophy is known as the *do-ocracy* because it empowers those who contribute effort to improve the hackerspace while removing elected or self-appointed leaders from the equation.

Noisebridge's wiki describes the philosophy thusly:

> *Do-ocracy: If you want something done, do it, but remember to be excellent to each other when doing so.*

"Be excellent to each other" is the primary rule of Noisebridge. A line pulled from the 1989 cult movie *Bill and Ted's Excellent Adventure*, it embodies the philosophy that a communal, nonhierarchal organization can accomplish its goals without the need for an excessive set of rules. As long as members are respectful of each other, the philosophy goes, members are free to do as they please. Of course, the do-ocracy is more of an ideal than a practical governing method.

See: https://www.noisebridge.net/wiki/Do-ocracy

ALT.PROJECT

LED MATRIX

Noisebridge's LED sign started off as a project thrown at a couple of members and ultimately became one of the most recognizable symbols of the space.

"The sign was one of the first things I hacked at when I got to Noisebridge," Josh Myer said. "My friend Andy handed it to me as a 'suggestion' for my first project, actually." Myer painstakingly reverse-engineered the board but ultimately got discouraged before his fellow member Nils McCarthy stepped up and finished the project. The sign originally was run by a built-in 386 computer connected to a socket with a missing chip; Myer and McCarthy wanted to use an Arduino to run it.

"After some quality time with an oscilloscope, I knew how fast I needed to shovel out data," Myer explained. "This turned out to be a bit problematic: the bit rate was over a megabit per second to refresh the screen! The Arduino could sort of drive it, but it wasn't nearly fast enough, so it yielded neat glitch effects instead of actually displaying anything."

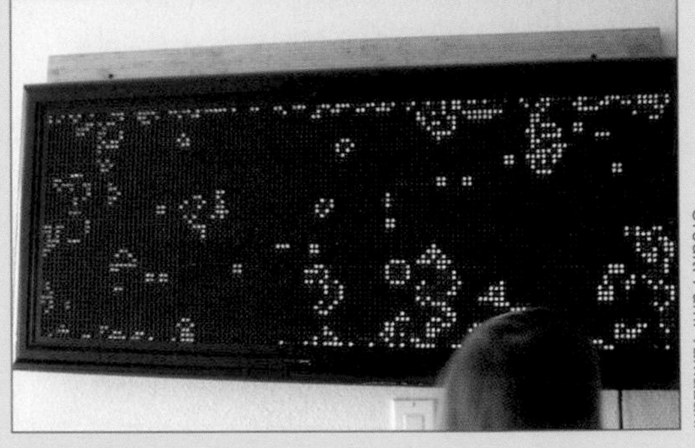

Credit: Fumi Yamazaki

▲ *A visitor checks out Noisebridge's impressive LED matrix set up to play Conway's Game of Life.*

That's when McCarthy took over the project. "He had the brilliant idea to use two Arduinos," Myer said. One Arduino controls the left half of the screen, and one controls the right half. "From there, he did a lot of hard work on getting the data out fast enough, coordinating between the two halves of the screen, and synchronizing with the refresh of the screen." The sign is 128 LEDs wide by 48 high, yielding over 6,000 pixels.

"I really enjoyed how the hackerspace took what was a half-abandoned project and breathed new life into it," Myer said. "I had done a lot of hard work and come to an impasse, and put off solving the new hard problem. And then someone with fresh energy comes along, sees the new hard problem, and runs with it. I believe that either of us could have finished the whole project, but having that handoff in the middle was really great."

CONWAY'S GAME OF LIFE

The LED screen currently plays the Game of Life, a simulation of cellular automation that was originally designed by mathematician John Conway in 1970. Each pixel equates to a biological cell and has one of two states: alive or dead. Each cell lives or dies based on how many neighbors (adjacent cells) it has. A live cell with less than two neighbors dies. A cell with two or three neighbors lives on to the next generation. A cell with four or more neighbors dies from overcrowding. The game is autonomous and runs algorithmically, without the interaction of human players.

Conway's Game of Life quickly became an obsession for mathematicians and computer scientists, who programmed simulations into some of the earliest computers. Because of this legacy, the game has become a favorite of hackers. One of the stable patterns of the game, the glider is even considered a symbol of hacker culture by many people. Author and open source advocate Eric S. Raymond proposed using the glider as the symbol of hacking, reasoning that hackers love the game and it was invented about the same time as the creation of the Unix programming language and the earliest nascence of the Internet.

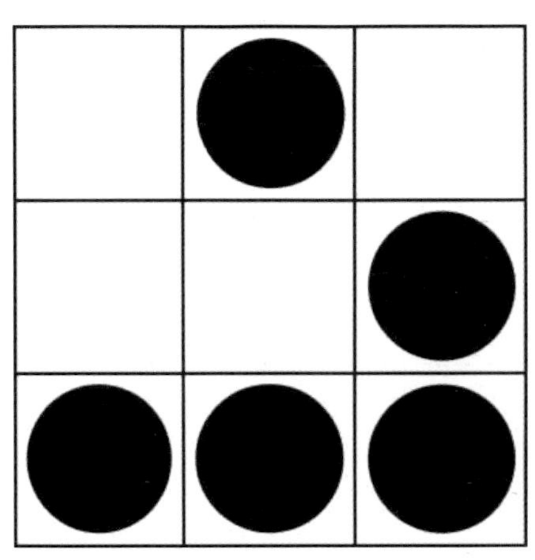

▲ *The glider, an object from Conway's Game of Life, has become a hacker symbol.*

Project page:

https://www.noisebridge.net/wiki/Big_led_screen

To see a video of the screen playing the Game of Life:

http://www.flickr.com/photos/mightyohm/4628002615/

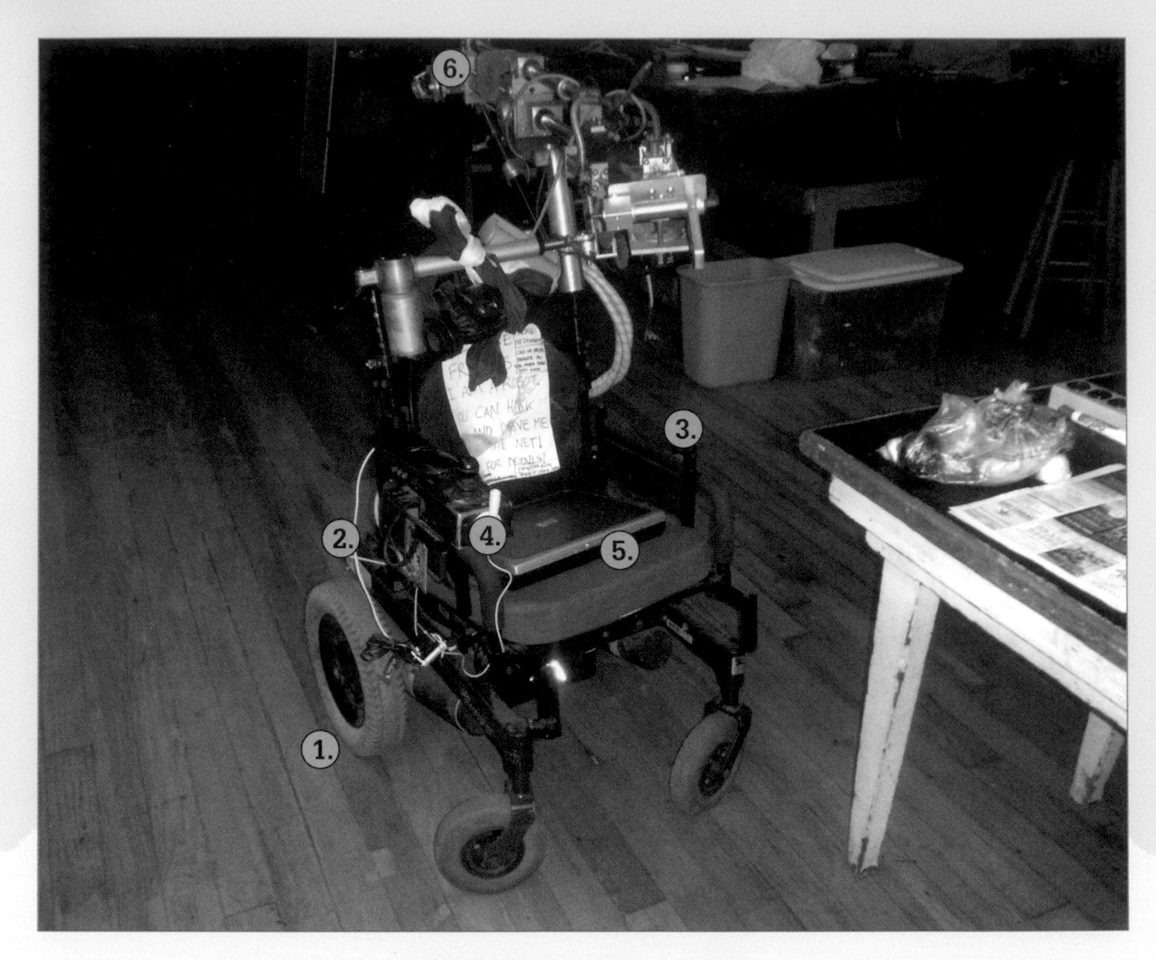

BUILD IT

While the project has evolved several times over the past couple of years, there are some consistent elements among all variations:

1. Get an electric wheelchair. The Noisebridge chair is an Action Arrow Storm Series.
2. Add the power supply manager. Most wheelchair users don't let their battery drain down all the way for long periods of time, which could potentially damage the car batteries that run the chair's motors. The team added an Arduino that shuts down the chair if the batteries charge down too low.
3. Add the interface board. This helps connect the chair to the computer.
4. Hack the joystick to add computer control.
5. Add a laptop running the custom Python scripts that the team developed. This allows people to log in to the computer and run it remotely from another computer.
6. Install optional extras like webcams, robotic arms, EEG control, and so on.

ADDITIONAL RESOURCES

Project site:

- https://www.noisebridge.net/wiki/Noise-Bot

Videos of the robot in action:

- http://www.youtube.com/watch?v=kJKXFENEDdE
- http://www.youtube.com/watch?v=R-5oosZ_wq4
- http://www.youtube.com/watch?v=XI_19brMqUU

Do It Yourself

Credit: John Baichtal

CRASH Space's mascot Sparkles is a My Little Pony with a soldering iron sticking out of its head, typifying the whimsy and cleverness of the hackerspace scene.

Why join a hackerspace? If you haven't found a hundred reasons in the preceding pages, the scene probably isn't for you. And it isn't for everyone. Hackerspaces are for those who yearn for friends to share their technological and intellectual passions, to help them build the projects of their dreams. Hackerspaces are for those who are energized by the work of others, who love to learn and share. If you don't like to learn, if you're afraid of others taking your ideas, if you have all the tools you want, the hackerspace environment may not be a good fit. If your reason for not joining a hackerspace is that there is none nearby to join, it's time to start your own! Here's how.

SPREAD THE WORD

You can't have a hackerspace without a core group of people to make it a reality. Depending on how much money you and your friends are willing to shell out, you can start a space with a mere handful of people. On the other hand, sometimes it's easier to spread the work and money around with a large group.

Those founding members are gold—in many spaces, they still form the heart of the organization years after their first meeting. Seattle's Jigsaw Renaissance was started in 2009 by seven friends. "All of them are still participants," co-founder Willow Brugh said. "From continued board members to holding occasional classes to using Jigsaw as their office."

In 2007, Jens Ohlig and Lars Weiler of Chaos Computer Club Cologne created the Hackerspace Design Patterns (http://www.scribd.com/doc/258372/Hacker-Space-Design-Patterns), which they presented at the 24th Chaos Communications Congress, a hacker convention in Germany. Their presentation and the accompanying slide deck have guided many hackerspace founders since then. Ohlig and Weiler suggested a minimum of 4 and ideally around 10 members to start with.

Here's a basic model: the fewer people you have, the faster the process will be, the more each member will have to pay, and the more the space will reflect your individual taste. By contrast, with a room full of people, the process will take longer as consensus is reached but each member won't have to pay as much.

Chances are, people willing to sit around in a meeting room or coffee shop discussing a

Credit: Pat Arneson

▲ *Hackerspace organizers meet at a coffee shop to plot their next move.*

Credit: Anne Petersen

▲ *Giveaways like buttons and stickers help spread word about the space.*

hypothetical hackerspace will be around long after the lease is signed. Ohlig and Weiler recommend recruiting people with strong personalities: "look for people who *have* authority (and get respect), not for people who *use* authority (and get laughed at)."

At this point you'll need to set up a web page, social media sites, wiki, mailing list, and forums to help spread the word. Just as importantly, be sure to complete a profile on hackerspaces.org, the worldwide aggregator of hackerspace information.

DECIDE ON AN ORGANIZATION TYPE

The central question when getting organized is should the hackerspace be for-profit or nonprofit? Let's examine the two types.

- For-profit

  In the United States this usually means a limited liability corporation (LLC) that basically protects the members from legal risk in the event that the business fails. Creditors might take the stuff in the space, but they can't go after individual members.

Credit: Pat Arneson

▲ *Prospective members of Twin Cities Maker vote to sign the lease to a new space.*

  Hackerspace LLCs are often begun by those with a particular vision of how they want their space to be run and don't want to lose that vision—to lose what they love about the space. This doesn't necessarily mean that the spaces are run as companies or that the person whose name is on the LLC acts as dictator. In many cases the differences between an LLC and a nonprofit hackerspace are invisible to the outside eye.

  LLCs are less concerned with hypothetically getting big donations because such gifts would not be tax deductible. On the other hand, they are extremely quick to set up as compared with a nonprofit, taking a couple hours on the Internet and a waiting period while the paperwork is processed. The LLC route is also appropriate if you actually do want to make a profit from the space. The "tech shop" phenomenon consists of all the trappings of a hackerspace but is focused more on offering tools to customers and less on fostering a community, and it's run as a business rather than a club.

  Some LLCs do not actively recruit members. While the majority of hackerspace groups are open to all and are constantly recruiting, some prefer to keep a small group of friends as a core, only recruiting when someone drops out.

- Nonprofit

  Despite the benefits of the LLC, most hackerspaces go the nonprofit route. The way this works is that the prospective nonprofit must arrange its management structure and finances in the way the government is expecting. In the United States, this means having a board of directors as well as a group of officers who (at least on paper) do the actual managing: a president, vice president, secretary, and treasurer. After a lengthy administrative process, the group might earn a federal nonprofit designation and not have to pay certain taxes, and donations to the group will be tax-exempt.

"A nonprofit business is a corporation which is prohibited by law and corporate documents from distributing profits and must reinvest any profits in the business," explained Twin Cities Maker development coordinator Michael Freiert. "This offers the next step of being able to receive tax deductible donations."

The board and officers imposed by law don't necessarily reflect the space's actual leadership structure. Often these public leaders do not attempt to exert their authority on the group; they're there to make the government happy. Instead, the group is run by consensus or anarchy or through some other governing structure.

Another disadvantage of the nonprofit is that it takes a long time to set up—sometimes upwards of a year—and the assistance of a lawyer. Often you can get a pro bono lawyer, but it is still a long and complicated process. In the meantime, some groups look for fiscal sponsorship through an umbrella organization. Space Federation (http://atrium.schoolfactory.org/spacefed/) is a 501(c)3 organization that accepts money from donors and passes it on to the member organization as a tax-deductible donation, with 10% of the funds kept to cover administrative costs.

Willow Brugh, a co-founder of Seattle's Jigsaw Renaissance, also helped create Space Federation. She sees it as being greater than just a vehicle for fiscal sponsorship. "By banding together, we also have a louder voice in larger culture and can start to lobby for insurance designations, appropriate zoning codes, and the like," she said. "With our powers combined, we can create a sustainable place in Mother Culture. By having each others' backs, we can ensure we uphold our shared values of transparency, mutual aid, adaptability, and emergence." The group is even working on a Make-a-Space Kit that will aid hackerspace founders in creating their new spaces.

Outside the United States, you'll have to find out how the laws for nonprofits and companies are set up in your country. You may find that your government actively supports small organizations. For instance, in Germany and Austria, political groups receive money from the government so hackerspaces in those countries often have a distinct political tinge to them that is typically absent from U.S. spaces.

NAME YOUR GROUP

One of the privileges of the founding members is coming up with a name for the group. Look at the hackerspaces in this book for examples of the ideas other groups have dreamed up: Tokyo Hackerspace, NYC Resistor, /tmp/lab, and Pumping Station:One. It can be anything you want it to be.

Credit: Jeff Keyzer

▲ *Hackerspaces.org is an international clearinghouse for hackerspace information.*

Usually—but not always—the name of the facility is the same as the name of the organization. In Minneapolis, two hackerspace groups—Twin Cities Maker and the Hack Factory of Minnesota—merged the groups to rent their new space. They named the organization Twin Cities Maker and called the facility the Hack Factory.

You should definitely choose a name you can live with, however, because it'll be very difficult to change the name later—usually it takes the dissolution of the organization. Some groups opt for a name that clearly spells out what the organization is about and where it is; others strive for a more unique or colorful name. "While 'Seattle Hackerspace' would certainly be more self-explanatory than 'Jigsaw Renaissance,' it just didn't work for us," said Brugh, one of the hackerspace's founders. "We wanted something which indicated tools, innovation, collaboration, and transdiscipline studies."

HOLD MEETINGS

To organize your group, you'll need to hold meetings—a lot of them. And the more people who are at these meetings, the longer it will take to reach consensus. Here are some tools that will help you get through these discussions.

- **Have an agenda**—While early gatherings are more about forging relationships, any working meeting needs to have an agenda. "Meetings should not be to discuss things; they should be to decide on things," Hack Factory founding president Mike Hord advised. "Long, drawn-out meetings can kill an organization, especially in the early stages where people are interested in getting to know one another." Having an agenda forces an end to contentious or drawn-out discussions.
- **Rules of order**—Many hackers have poor social skills and like to interrupt each other. Even worse, some toxic individuals seek to take over meetings out of a love for pedantry. The solution is a set of guidelines for managing meetings called *rules of order* (http://en.wikipedia.org/wiki/Rules_of_order). The rules call for an ordered, polite series of verbal exchanges whereby the majority opinion of the group is decided upon. The system most widely used is called Robert's Rules of Order, which lays down specific procedures for motions, amendments, and votes. "These are important to keeping a process moving forward," Hord explained. "One person cannot deadlock the meeting if everyone else—or even 'just' a majority—is in agreement."

  You don't have to use rules of order in every meeting, but at very least there should be an acknowledged chairperson of the meeting who can impose the rules if arguments break out.
- **Keep minutes**—It's important to keep minutes of every meeting, if only to inform people who missed the meeting what happened. More importantly, most hackerspaces are very open and sharing organizations, and any sort of deliberate secrecy is frowned on. Having a paper trail in the form of meeting minutes encourages that environment. "Transparency and member involvement are key to the success of a hackerspace," Hord agreed. "Pre-posting an agenda as well as posting minutes allows people to involve themselves in issues that they have a key interest in and prevents them being blindsided by decisions that affect them."

  Of course, a high percentage of members will never read them, but the important thing is that they have access to the hackerspace's records if they're interested.

Additionally, groups interested in attaining nonprofit status must keep minutes and other administrative minutia to stay in the good graces of state and federal regulations.

- **Reach consensus**—Many hackerspaces place a high value on decision-making via consensus, rather than having the leadership make the decision. Usually this involves discussing the matter until all present are in agreement. "My general rule is that you are only allowed to dissent if you have another solution you're willing to act on," Brugh said. "That said, everyone should have a chance to be vocal about where they stand—vocal with a time limit."

  Hord agreed. "Even after [a matter] has been moved, seconded, and voted upon, some people will still insist on attempting to reengage," he said. "Parliamentary procedure suggests gavel pounding and declarations of 'out of order,' but in a less formal setting, a firm but polite 'we're done discussing that,' repeated over and over will usually suffice."

- **When in doubt, call bike shed**—There is a recurring phenomenon in collective groups like hackerspaces called *bike shed* (http://en.wikipedia.org/wiki/Bike_shed), in which a trivial matter can derail an important undertaking. The classic example has a group agreeing to build a shed to store bikes but spends all its time arguing over what color to paint the new shed and end up not building it. One of the arts of meeting facilitation is learning to recognize bike shed behaviors and put a stop to them.

  "Consensus building," Hord explained, "is about knowing when to apologize and say, 'Sorry, but you aren't going to get your way this time because the detail you want to argue about isn't important enough to derail the entire project.'"

CREATE BYLAWS AND RULES

The rules under which the typical hackerspace operates can be seen as a spin on the "hacker ethic" (http://en.wikipedia.org/wiki/Hacker_ethic) that has evolved over the years: the sharing of tools and knowledge; decentralized leadership structures; and desire to move past racial, gender, age, and other stereotypes not related to one's skills. That said, it's good to have the rules and bylaws clearly laid out.

Bylaws are the rules under which the organization functions. They govern such matters as officer term length and how often board meetings are held. QC Co-Lab's bylaws (http://wiki.qccolab.com/index.php?title=By-Laws) are typical in this regard. The average set of bylaws are dry stuff, apropos mainly for "board-level" organization of the group. That said, you should definitely use the bylaws to spell out the extent of the powers of the elected leaders. Beginning with a benign president who doesn't abuse his or her powers doesn't mean the second president will be as cool.

Rules, on the other hand, focus on more practical matters as wood shop safety and where projects get stored. Although each set of rules will necessarily be unique to the space and its founders, certain patterns become evident. For instance, CCCKC's rules (downloadable as part of the membership packet at http://www.c3kc.org/membership) work as a Top 10 list, though starting at 0—traditional in hacker circles. CCCKC's rules read as follows:

0. Be nice to each other.
1. Be careful, but not too careful.
2. Respect everyone's equipment.
3. All members of the press must identify themselves upon entry.
4. Only take pictures with the subject's permission.
5. No solids in the sink.
6. No non-members are allowed without being accompanied by a member.
7. Everyone using tools must sign the paperwork.
8. Pick up after yourself.
9. Turn off the lights when you leave.

Hackerspace rules typically boil down to three: be courteous, be safe, and be clean. You definitely don't want a long and complicated set of rules because no one will read them. Indeed, Noisebridge has only one: "Be excellent to each other" (https://www.noisebridge.net/wiki/Noisebridge_Vision), a reference to the cult movie *Bill & Ted's Excellent Adventure*. Noisebridge's philosophy can be expanded to encompass several of CCCKC's rules. For instance, if you don't pick up after yourself, you're not being excellent to the person who follows.

One area that should be clear in the rules is that the spirit and letter of all laws shall be respected within the space. Many people still associate hackers with lawbreaking, and the rehabilitation of the term—not to mention the success of the organization—depends on members being above reproach when it comes to the law.

Ultimately your group will have to come up with its own rules. They might look very different than those previously mentioned if you are in a unique situation or if the founding members have a particular pet peeve they want to combat.

CHOOSE LEADERS

Whether or not to have official leaders, how to choose them, and how much power to give them are all questions that many hackerspaces struggle to resolve. The wonderful thing about the scene is that it seems as if every space comes up with its own solution. Here are some options:

- **Formal leaders**—Some spaces have no formal leaders, but many do. Indeed, as discussed previously, the government might even expect you to have a specific leadership structure. Some hackerspaces like the classic president-VP-secretary-treasurer format and stick with it. Formal leadership structures typically call for the board to make all decisions, only occasionally opening up the floor for regular members to weigh in.

Credit: Paul Sobczak

▲ *Board meetings, like this one at the Hack Factory, can be where all the decisions are made—depending on the hackerspace.*

However, many hackers find that structure very limiting and contrary to their idea of how the space ought to be run. For instance, it doesn't have a place for hackerspace-critical leaders like a network manager or wood shop foreperson. The authoritarian model also can cause a degree of paralysis, where non-board members won't take the initiative on a necessary hackerspace task for fear of "getting in trouble."

If you're going this route, be sure to build in a process for removing abusive or incompetent leaders into the bylaws, and remember that the final responsibility of the leader is to step aside in an orderly and timely fashion when he can no longer perform the job's duties.

- **Informal leaders**—Ohlig's and Weiler's design patterns urge the use of temporary leadership to solve problems. Arguably some of this can come from European spaces' counterculture atmosphere, as well as the hugely influential Noisebridge. Their consensus process (https://www.noisebridge.net/wiki/Consensus_Process) is widely considered to be the gold standard for non-hierarchical management, though it might not be right for your group.

  For example, the space is a pigsty and needs to be cleaned. One member gets mad and stands up and starts ordering people to collect recycling, sweep the floor, and organize the tools. It happens, and then the ad-hoc leader sits down and works on his project.

  If a couple of members want to move a work table to another room, they just do it. This philosophy is called the do-ocracy (https://www.noisebridge.net/wiki/Do-ocracy) a method of non-hierarchical leadership advocated by Noisebridge and followed by many other hackerspaces. In the do-ocracy, whomever wants to take the effort to do something (in this example, move the table) goes ahead and does it, and if someone objects, the table-movers are obligated to undo what they did. Complicated and important matters are discussed in a weekly meeting until consensus is reached.

RAISE MONEY

As tacky as it sounds, it's a reality: a continual flow of cash is the number-one requirement for a healthy hackerspace. Let's be realistic—maintaining a lease and paying the electric bills, insurance, and other expenses

Credit: Jeff Keyzer

▲ *Need to make money for the group? Sell stuff!*

Credit: Anne Petersen

▲ *Dues are the single most important source of income to most hackerspaces.*

keeps the padlocks off the doors and lets members hack. Here are some tips for managing money:

- **Elect a totalitarian treasurer**—Ohlig and Weiler's Design Patterns say "Elect a totalitarian treasurer," and it's true. If a space wants to survive, it has to have a person—the treasurer—whose job is to collect dues, even if his or her own friends are the deadbeats. The ideal treasurer won't take excuses and won't make exceptions.
- **Set a budget**—In addition to collecting money, the treasurer needs to manage the budgets professionally. The big items—rent, insurance, water, trash, and heat—are obvious. But don't neglect seemingly optional supplies like cleaning products and toilet paper. Definitely budget for a robust Internet package; nothing makes a room full of hackers crabby like no tubes.
- **Solicit donations for unexpected expenses**—Often, expenses that weren't budgeted for end up getting paid ad-hoc by members, but you shouldn't count on this generosity. One solution for paying for these items is to solicit donations from members who use the relevant area. For instance, in one hackerspace, the MakerBot users realized they needed a new power supply for their 3D printer. It was decided that the people who used the machine should chip in to buy the new part.
- **Diversify income stream**—Dues are the number-one source of income for most hackerspaces, but it shouldn't be the only one. Sometimes the ebb and flow of interest leaves the rolls a little thin, and you shouldn't have to worry about making rent just because a couple of members are late on their dues. Other sources could include class fees, beverage sales, rent parties, t-shirt sales, and admission fees to special events. Be aware of your local laws governing raffles and other moneymaking operations.
- **Solicit donations**—Donations, whether on the books as tax-deductible or otherwise, are a reality of the average hackerspace. Most of the time it takes the form of a few bucks dropped into a donation jar or a member bringing in her power tool to keep at the space. Sometimes bigger guns are needed, and that's where microdonation sites like Kickstarter come in.

When he started his new hackerspace, Nick Farr set up a Kickstarter page to solicit donations. He asked the space's keyholders to contact their friends and family to donate a total of $3,000 to help get the space set up. They ended up receiving $5,000. "I guess that's the key," Farr said. "Actually reaching out to your network in ways that go beyond social media. We all dropped a few key emails

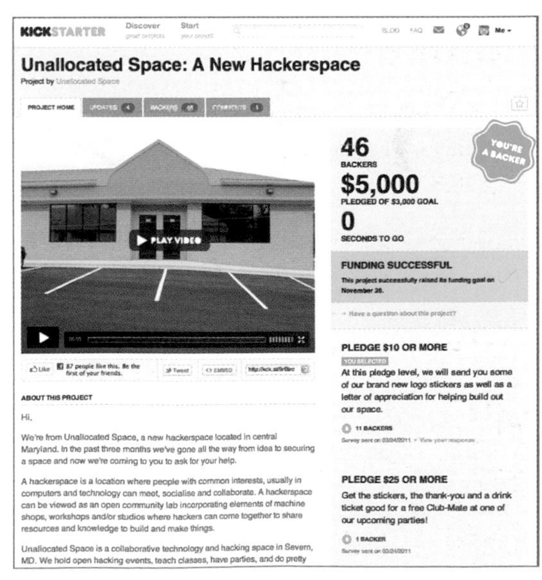

▲ *Unallocated space received $5,000 in donations to offset renovations.*

to our friends and people we've supported with their endeavors, explaining what our hopes and dreams were for the space, and they responded well beyond our expectations!"

- **Be financially independent**—A note on sponsors and financial benefactors: Don't do it. You should never be beholden to an organization or a person for the survival of your group. If you need such a group, you probably need to recruit more members. If you absolutely must accept the help of a benefactor, make independence your first priority. At the same time, resist the temptation to use someone's house or apartment, or to save money by letting someone live at the space. Anytime you combine someone's work area with someone's sleeping area, there will be conflicts.
- **Save for a rainy day**—Just like your home budget, you should save a little extra when you're flush so that you won't be scrambling for cash during a slow period. Wouldn't it be nice to have three months of expenses in the bank?

Credit: Paul Sobczak

▲ *Want a metal shop in your new space? Plan accordingly when looking at properties.*

FIND A SPACE

Look how far we are into this chapter, and we're only now talking about finding a space! That demonstrates how much work goes into getting organized enough to shell out for a lease. Unless you're lucky, however, finding the right space can consume just as much time. To be clear, you'll want to find the right space from the beginning because it's very expensive and labor-intensive to switch facilities. More than one hackerspace has disbanded after losing its lease. But with that much at stake, how can you be sure to find the right space?

Some groups come by free space, either contributed by a member or by an umbrella organization. The problem with this is that the group will always be beholden to this benefactor and, at best, will live with the possibility of being abruptly evicted or having the parent group set rules for activities that can take place in the space. It's better to be independent. Even subleases offer problems—Twin Cities Maker in Minneapolis nearly found themselves homeless when the company from whom they'd subleased abruptly moved out.

One very important factor is to find a space with the right kind of electricity. Can the space support the power needs of a laser cutter, arc welder, and kiln? If not, you might want to look for another facility or be prepared to upgrade the power yourself.

Credit: Anne Petersen

▲ *Not every facility—or landlord—is ready to accommodate a chemistry lab. If you want one, be sure to keep it in mind while hunting for the perfect space.*

Jigsaw Renaissance's Willow Brugh offers some advice on finding the perfect facility. "Your landlord is the most important thing," she said. "Either a benevolent absentee landlord or one who *gets it* are best." You'll want a landlord who doesn't care if there are soldering irons, welders, or laser cutters being used in his property. Obviously, some of the trappings of a hackerspace can be pretty intimidating to someone from outside the scene. On the other hand, many hackerspaces are willing to move into a run-down property and fix it up, and that generates a lot of goodwill during lease signing.

The most kindly landlord won't help if the space isn't right, however. The first thing you should do when looking at properties is to keep in mind the activities the members would like to undertake at the space. If there are a lot of foodies in your group, for example, be sure to find a space with a kitchen. Milwaukee Makerspace's members are interested in car projects and found a space with a drive-in work area to facilitate this kind of work.

Ohlig and Weiler's Design Patterns suggest a lot of small rooms rather than one big space. This allows members to explore many areas of interest that might conflict with others. For instance, a wood shop generates a lot of sawdust, so you don't want it next to your server room. A chemistry area might have fumes, so it'll need a tight-shutting door and a vent to the outside. You'll definitely want some sort of chill-out area or lounge, with snacks and beverages available for purchase, along with a lockable closet for storing cases of pop and other consumables.

To a degree, choosing the right neighborhood can be almost as important as choosing the right facility. You want a neighborhood with lots of parking, coffee shops, takeout, and other services. The Hack Factory in Minneapolis has a classic full-service hardware store less than a mile away and sends a lot of business their way. But having great neighbors can also be important. Hackers tend to work through the night and often experiment with chemicals, vehicles, and pyrotechnics, so avoid quiet and cozy neighborhoods. Ohlig and Weiler's Design Patterns agree: "As hackers, you do not live the majority lifestyle—look for neighbors who are also weird and outside the majority."

EQUIP THE SPACE

Hackerspaces accumulate a lot of stuff, ranging from table saws to heaps of useless junk. The important thing is to have the right tools and furnishings. Too few or the wrong equipment and members won't be able to do the activities they want. Too much equipment can be a detriment as well. Imagine a wood shop so full of tools brought in willy-nilly from members' garages and storage lockers that no one can get anything done—you probably don't need four band saws at the space. Here are some common areas found in hackerspaces:

Credit: Paul Sobczak

▲ *A fully equipped craft area is an asset to any space.*

Credit: Mitch Allman

▲ *Members of Ontario's Kwartzlab relax in their space's chill-out area.*

- **Chill-Out room**—Don't underestimate the importance of a lounge. Many members join mainly to socialize with like-minded hacker types, and you don't necessarily want them standing around the metal shop. Stock your chill-out room with couches, chairs, recliners, coffee tables, incandescent lights, stereo equipment, and video games. You might want to consider a soda machine because these can be an income stream as well as an amenity.
- **Crafts area**—The catch-all term for knitting, sewing, leather working, and so on. Crafters need a lot of clean table space and good lighting. You should invest in an industrial sewing machine, an electric kiln, craft scissors, hot glue guns, a vinyl cutter, paints, a screen printing setup, and so on. The Hack Factory has a mini photography area with umbrella lights and a white backdrop.
- **Network operations center (NOC)**—Every space needs networking equipment, even if it's just a cable modem. Equip your NOC with racks, backup power, network routers, wireless routers, and servers. You might want to lock the room so self-styled experts don't mess with the setup on some Saturday night when the Internet is acting up. If you've designed your own security system, its back end can live in the NOC.
- **Fabrication lab**—Many hackers are interested in rapid prototyping, and having a fablab available to them will help them build their dream projects. Fabrication tools cost many thousands of dollars, but many spaces do what it takes to buy or build them. Most groups have at least one MakerBot or similar 3D printer, and many also have a laser cutter. Equip your lab with those two if you can, and add CNC mills, 3D scanners, plasma cutters, and other rapid prototyping machines.

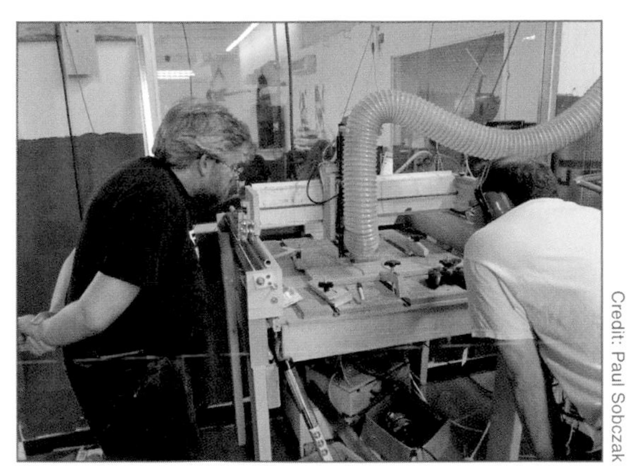

Credit: Paul Sobczak

▲ *Every hackerspace needs rapid prototyping equipment like this CNC mill.*

- **Wood shop**—A hackerspace woodworking area can take you back in time to shop class, but it's a critical ingredient. You should stock your table saws, band saws, drill presses, routers, grinders, and an air compressor with tool attachments. Hand tools, too, are a necessity. Wood shops use a lot of consumables like saw blades, sandpaper, and drill bits, so stock up on those. Also, be sure to get at least one shopvac for cleaning up sawdust. Don't neglect safety gear like goggles and earplugs.

Credit: John Baichtal

▲ *Most hackerspaces have at least one drill press.*

Credit: Paul Sobczak

▲ *Brewing classes are among the many educational offerings of hackerspaces.*

- **Classroom**—Most hackerspaces hold classes, so having a separate classroom ensures that people can still work while a session is underway. In addition to comfortable chairs and work surfaces, a decent sound system and a projector are good bets.
- **Electronics shop**—This is one of the most popular areas in many hackerspaces, so much so that many don't have a dedicated room for electronics, and instead merge it with the social space. You'll need multimeters and oscilloscopes, as well as solder reflow ovens and PCB etching stations. Get a lot of soldering irons for your electronics area because soldering classes always draw a big crowd.
- **Metal shop**—A metal shop can be a little obnoxious, with all sorts of sensory overflow, like light from arc welders, noise from grinders, and fumes from oxy-acetylene torches. However, you'll be happy to have a full-fledged metal shop in your space. In addition to welders, be sure to get grinders, a plasma cutter, metal-cutting band saws, safety gear, hand tools, a metal-casting furnace, and an anvil.

Credit: Paul Sobczak

▲ *Every hackerspace needs a metal shop.*

- **Storage**—Be sure to reserve room to put stuff because a bunch of hackers generate a lot of projects (and half-built projects) that will need storage. One solution is to rent storage space to members. For instance, $5 a month for a lockable cabinet is a good deal and puts money in the coffers. Even better, it ensures the contents of the locker don't get left behind when a member quits; orphaned projects are a big problem in hackerspaces.

In the end, you should let your members drive the tools. Your founding group may like one thing, but adding 20 new members with diverse interests is likely to expand your hackerspace's coverage in ways that surprise you.

Credit: John Baichtal

▲ *Storage and safe-keeping of projects, tools, and materials is a continual struggle for many hackerspaces.*

MAKE IMPROVEMENTS

If the perfect space simply can't be found, one solution is for hackerspace members to remodel the place themselves. After all, you have dozens of people who are handy with tools and motivated to improve the facility, so why not use them? Often members will have friends who are licensed plumbers and electricians, lending critical support to help make the project happen and making the landlord happy by improving his building for free.

Credit: C-P

▲ *Unallocated Space's loft adds much-needed square footage to the space.*

Credit: C-P

▲ *The hackerspace's new "chill-out area" overlooks the main workshop.*

In D.C.'s Unallocated Space, the members took advantage of a high ceiling to add a loft. Their treasurer, Nick Farr, described the project: "Soon after we moved in, one of the corners ended up becoming our lounge/chill-out area where our big TV, video game systems, and couches were. After I found a used steel mezzanine from a surplus place in Rhode Island that would fit that corner, we decided to 'elevate' our lounge area, to leave the rest of the shop floor as workshop space." The group ended up building the loft by hand, "which was a really dumb idea," Farr said. "We raised the 500 lb. stairwell into place using 2×6s and pallets—which is not something I would recommend. But we're really, really happy to have a space 'above it all' to chill out."

Other possible upgrades include painting, building shelves, ripping out old carpeting, and dividing up large rooms into small ones.

KEEP IT CLEAN

Hackerspaces tend to be messy, filled with half-finished projects, junk for salvage, and tools that haven't been put away. Never mind soda cans, solder snippets, disorganized shelves, overflowing trash cans, sawdust, and general messes and spills. There are coffee cups stacked in the sink and a bathroom that hasn't been scrubbed down in weeks. It's very easy to let the space get horrid if no one steps up to clean and organize. Here are some tips for keeping a tidy space:

- **Quarterly deep clean**—No matter what, chances are the hackerspace will need to be thoroughly cleaned every few months. Find the member who is the most irate about the current mess and put that person in charge of the deep clean. Hackers are often willing to clean but don't want to put any thought into it, and having someone in charge of the operation will help motivate them.

- **Load up on the cleaning supplies**—Stock a broom closet with cleaning products, brooms, rags, scrub brushes, vacuums, and so on. Don't let the space get messy for lack of supplies.
- **Clean after events**—Organizers of classes, talks, and other gatherings in the space should schedule 15 minutes at the end of their events for attendees to clean up their area after the event is done.
- **Periodic junk cleanout**—Found and donated junk tends to pile up. Every few months, go through the junk and weed out the stuff that has been sitting on the shelves for a while. Find out how to recycle computers and other electronics in your area, and dispose of the junk responsibly.

SECURE THE SPACE

A hackerspace can contain many thousands of dollars in tools and equipment, and it's definitely suggested that you get some sort of alarm system and modern locks. The need to secure the space has given rise to the phenomenon of the keyholder, a member of good standing who is trusted with a key to the space. Simply put, the hassle of recovering a key from a former member, not to mention the expense of cutting new keys, means that these spaces are hesitant to allow every member to hold keys. (In Forskningsavdelningen in Malmö, Sweden—featured in Project #13—new members have to learn how to cut their own keys on a key-making machine.)

In the last few years, more and more spaces have been building their own security systems (see Project #19 for an example) that utilize RFID or numeric entry codes, allowing every member, in effect, to be a keyholder. The advantage of these systems is that a former member's key (or code) can simply be deactivated, reducing the number of keys floating around.

Credit: Deven Fore

▲ *TheTransistor's homebrewed security system admits members when the right code is entered.*

BE SAFE

Credit: John Baichtal

▲ *Seriously, be safe!*

With a building full of power tools, lasers, flammable gas, and so on, it's absolutely important to maintain a strict safety policy. First, supply members with whatever skin, ear, eye, and respiratory protection they'll need. This means large stocks of goggles, earplugs, welding masks, gloves, lab coats, and so on. Post warning signs by potentially lethal tools like band saws and lathes, and be sure the appropriate areas are equipped with fire alarms, extinguishers, and a robust first aid kit. If chemicals are in use, look into installing an eyewash station.

Member training is important as well. Hold classes to certify new members on the dangerous and fragile tools. Educate members on safe clothing and attire, and set sensible rules for tool use, like asking members not to use certain tools if they're by themselves in the hackerspace. No one wants to get injured and not be able to get help immediately. Finally, you might even want to appoint a safety coordinator whose job is to set safety rules and conduct training sessions.

RESOLVE CONFLICTS

As you establish your hackerspace, you'll probably go through a honeymoon period where everyone is giddy with the new clubhouse. Inevitably, however, conflicts will arise. Someone might show up at the space to find his or her favorite area rearranged. A rebellious member might not like a new rule or policy laid down by the board. Or two hackers might simply not get along.

"Defining your values helps prevent conflict, too," explained Brugh. "Saying, 'we stand for these things, if you don't adhere to these, you can't play' makes it much easier for everyone to know how to interact. Jigsaw has a waiver like everyone, but we also have a membership agreement and safe-space agreement."

The best way to deal with conflicts is to sit down and talk it through, with a moderator if necessary. Other methods could include having a respected but uninvolved person talk to the troublemaker(s) in private. If that doesn't work, an all-members' meeting may do the trick: discuss the matter until a resolution is found.

Sometimes the problem is that someone cares *too* much. Inevitably, there will be unemployed hackers with insane amounts of time to devote to the space. These folks are both an asset and a hindrance. Although they have a lot of time and energy for the organization, they become obnoxious and abuse the do-ocracy; when everyone else is at work, they're hanging out at the space making it more like they want, annoying their fellows with more stringent schedules. Learn how to cool these members down without alienating them.

Finally, have a mechanism in the bylaws for expelling members, but don't use it unless you must. Sometimes, a hacker is too toxic to be salvaged. Maybe he steals, acts in a violent or unsafe manner, or doesn't play well with others. If that is the case, there should be a procedure for expelling the member.

SHARE EVERYTHING

Every space is different, and every organization has the right to choose for itself what sort of information it wants to share, but most hackerspaces share everything from tool lists to current projects. The culture of learning and doing, many hackers believe, thrives on people sharing what they've learned.

When you begin forming your group, you should have a wiki, forums, a blog, photo hosting, an IRC channel, and social networking sites like Facebook and Twitter. This stuff is obvious, but don't overlook more creative ways of sharing what you do.

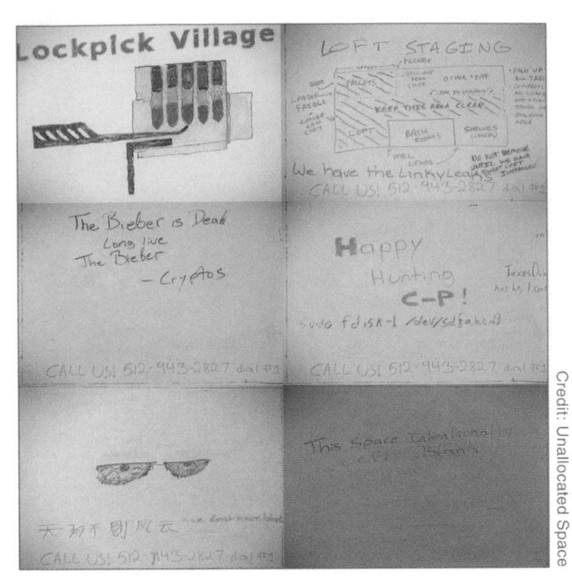

Credit: Unallocated Space

▲ *Several snapshots of Unallocated Space's white board; every time the space is shut down for the night, another photo is snapped.*

For instance, Unallocated Space in Severn, Maryland, has a camera focused on their white board (http://www.unallocatedspace.org/thewall/), and every time the security system is engaged—for example, when the space is closed for the night—the camera shoots a photo of the board and uploads it to the Web.

This sharing helps casual members keep abreast of what's going on in the space, it lures in potential members who want to be part of the fun, and it inspires other hackerspaces to attempt projects and activities as cool as Unallocated Space's.

RECRUIT NEW MEMBERS

Most hackerspaces are in continual recruitment mode, simply because of the importance of dues to keeping the organization alive. But how does one find these recruits? One trick is to lure visitors to the space with fun and engaging events—for example, classes, talks, and parties.

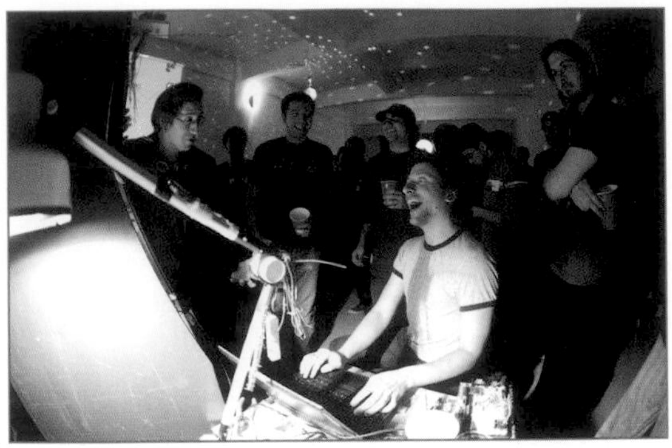

Credit: Bre Pettis

▲ *One of NYC Resistor's rent parties attracts a crowd of visitors to the hackerspace.*

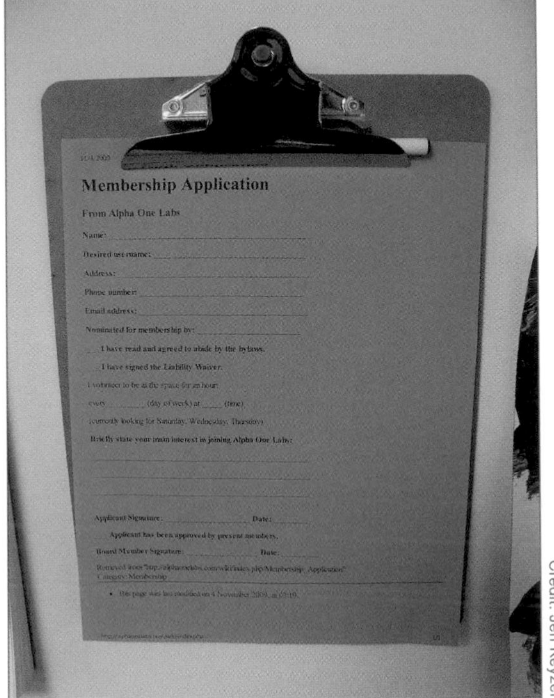

Membership Application

From Alpha One Labs

Name:

Desired username:

Address:

Phone number:

Email address:

Nominated for membership by:

I have read and agreed to abide by the bylaws.

I have signed the Liability Waiver.

Briefly state your main interest in joining Alpha One Labs:

Applicant Signature: Date:

Applicant has been approved by present members.

Board Member Signature: Date:

Credit: Jeff Keyzer

▲ *Alpha One Labs' sign-up sheets are ready for any prospective members who stop by.*

Probably the most useful recruitment tool is the regular open houses that most hackerspaces hold. Often they're on a Tuesday or Wednesday, and most of the members of the space congregate to socialize and work on projects. Visitors who attend can get a taste of the atmosphere at the space and meet the other members. Be sure to have a stack of business cards, flyers, and sign-up forms available, and make sure someone's able to accept dues and a key deposit to take advantage of impulse sign-ups.

You'll definitely want people of both genders and all races and backgrounds. It isn't political correctness; it's simply a fact that these people can bring a vastly different perspective and set of skills to the table, and the more knowledge your hackerspace has access to, the stronger and more useful it will be to members. That said, don't try to recruit anyone and everyone. If the space isn't right for them, getting to pay dues for a month or two won't help anyone—you want someone who'll stay for years. "Be what you say you are," Brugh agreed. "If you're an event space, hold solid events. If you're a hardware hacking space, hack some hardware. It's basically walking the walk but also talking about it."

You can also recruit members outside the space. Fairs and conventions are a great way to showcase the group's projects and interact with potential members. Many hacker-favorite gatherings like Hackers on Planet Earth and Maker Faire have "hackerspace villages" where members congregate.

TEACH CLASSES

Hackerspaces are learning places, and oftentimes that learning takes place in the form of a class. Topics can include anything from woodworking to programming to electronics: any subject a member is good at and is willing to teach.

In addition to educating people, hackerspace classes are good in two ways. First, they attract a group of people who aren't necessarily interested in membership but want to learn the topic. Just getting them in the door is a victory because it exposes them to the hackerspace environment, so some of them might eventually become members. Another benefit is the money classes bring in. Charging $20 to attend a class might not seem like a lot of money, but it adds up. Many spaces end up deriving a significant portion of their operating expenses, using the money to fund improvements in the area related to the class. The proceeds for the welding class, for instance, can go to supplies for the metal shop, encouraging members who use the area a lot to volunteer to teach a class.

Credit: Brant Holeman

▲ *Members of Milwaukee Makerspace attend a class on Arduinos.*

As another inducement, many spaces kick back some of the proceeds to the teachers to offset their dues, so an underemployed hacker can teach rather than pay—and everyone wins.

HELP THE COMMUNITY

Could hackerspaces be an important resource during disasters? It may be that hackers' unique mix of skills make them ideal for ad-hoc support for the local population. Akiba, the hacker from Tokyo Hackerspace (profiled in Project #3), explained that our ability to

Credit: Tokmyo Hackerspace

▲ *A kimono lantern, an easy-to-build solar light kit, can be distributed during power outages.*

purchase what we need breaks down, making the ability to create, repair, and modify extremely important. "You suddenly need a lot of things customized to a particular situation," he said.

During the 2011 Tokyo earthquake, Akiba and his hackerspace friends responded to dangerous radiation levels by hooking up Geiger counters to the Web. "Geiger counters were devices that used to be only purchased by people working with radioactive materials, paranoid militia, and weather geeks. Now they are becoming an everyday item in Tokyo."

"Hackerspaces have tools and members with knowledge about how to make things from scratch and modify devices to suit their needs," explained Akiba. "If a major environmental disaster were to hit San Francisco, I can imagine that Noisebridge would have weather balloons, monitoring stations, and software to disseminate all the information to the public within days."

Hackerspaces can also create ad-hoc networks to help each other. When Akiba looked for help creating kimono lanterns, solar-powered emergency lights, he was amazed at the response. "Within one day, we had offers coming out of hackerspaces in Oklahoma, Arizona, Detroit, Hong Kong, San Francisco, Germany, Singapore, and many other places."

NOW WHAT?

It's a lot of work and can be a big headache, but building a hackerspace can also be incredibly rewarding. It's hugely inspirational collaborating with so many brilliant people with interests different from your own. You can build the projects of your dreams for less than a cable bill, learn skills you never knew existed, and find companionship in smart people with overlapping interests. Years from now, when hackers from the other side of the world have heard of the accomplishments of your space, when dozens of talented and adventurous hackers have found a home there, you can fondly look back on your first hesitant steps.

Where your group—or the scene in general—will be in 20 years is hard to predict. The first wave of hackerspaces that were formed in the United States in the early '90s were very different from the groups we know today. Europeans took the concept and put their own spin on it, transforming the secretive hackerspace into an open, political, almost countercultural movement. The third wave of hackerspaces that began with the founding of Noisebridge, NYC Resistor, and a few others is still going. It has moved past the U.S.-Europe dichotomy to include every continent on Earth. It has largely dropped the secretiveness of the first wave and the overt politics of the second and learned from both of them.

The hackerspace of the future might take on the role of a community resource. Hackers' tinkering focus and the knowledge of tools and technology potentially make them invaluable in circumstances where the local community can't help. "In normal life, you can pretty much buy anything that you would need for everyday living," Akiba explained. "But in a disaster scenario, this all breaks down. At that time, normal life ceases and you suddenly need a lot of things customized to a particular situation." All of a sudden, hackers' ability to prototype new devices and love of outmoded skills like knitting and brewing make them incredibly useful. Many hackers are ham radio enthusiasts, and being able to operate communications networks during disaster situations can be critical.

Spaces can be business incubators, giving entrepreneurs the tools and space they need to develop products. Not needing to rent space or buy tools lowers barriers and lets more people have the chance to start their own business.

Or hackerspaces can look more like schools. Why should impoverished school districts maintain wood and metal shops when they could theoretically hold class in the hackerspace? With these groups focused so much on education, it wouldn't be a stretch for those same rapid prototyping, metalworking, and knitting classes to serve schoolchildren as well as adults.

Looking at the history of the movement, it's incredible how new and exciting it all is. Spaces founded a mere four years ago are considered established and respected, while most groups have been around less than two years. Who is to say what the future holds?

We're at the very forefront of a new and exciting phenomenon. Come be a part of it.

Hackerspeak

Like any subgroup, hackerspace members use a specialized vocabulary. The following glossary explains some of the more obscure terms found in this book.

1337: See *elite*.

***2600*:** Legendary hacker magazine, published quarterly since 1984.

31337: See *elite*.

3D printer: A printer that lays down layers of plastic rather than using ink. As the plastic builds up, a three-dimensional object takes form.

arduino: An easy-to-program microcontroller, favored by everyone from artists to electrical engineers for its ability to rapidly bring an idea to the prototype stage. Arduino is also the name of the language that programs arduinos.

bike shed: In a hackerspace or other communally run organization, *bike shed* signifies a discussion that has been derailed by arguments over petty issues. Basically, the archetypal bike shed doesn't get built because people are arguing over what color to paint it.

circuit bending: Altering the electronic circuitry of a noise- or music-generating toy to achieve interesting audio effects.

Club-Mate: The unofficial hacker beverage of choice. It's made from yerba-maté leaves, is heavily carbonated, and is definitely an acquired taste.

CNC tool: *CNC* stands for computer numerically controlled, meaning that the computer tells the tool where to go, resulting in extremely accurate automated cutting of wood, plastic, and metal.

Conway's Game of Life: A game that simulates cellular life cycles, created by mathematician John Horton Conway and first published in *Scientific American* in October 1970.

DefCon: The premiere hacker convention, focused primarily on computer security issues but embracing the full spectrum of hacker interests including lock-picking, electronics, and more.

do-ocracy: Leadership by those with the time and energy to act on their ideas. You want to put that table there? Do it; it's a do-ocracy.

elite: The quality of hackerly excellence, often abbreviated *'leet*, *133t*, *31337*, and so on.

first-wave hackerspace: Private spaces created in the '90s as clubhouses for small circles of friends.

g-code: Instructions for a CNC device. For instance, "move the toolhead 10mm on the Z-axis."

hackathon: All-night session of coding, soldering, knitting, sawing, and so on.

Hackers on a Plane: A fabled 2007 plane trip credited for having inspired the third wave of hackerspaces.

hackerspace: A DIY-friendly, cooperative workshop that inspires and educates its members.

handle: An alias, similar to a screen name, traditionally used by hackers.

H.O.P.E.: Hackers on Planet Earth, a long-standing hacker convention, held in even years in New York City.

Instructables: DIY-favorite how-to site specializes in providing step-by-step instructions for many different projects.

laser cutter: An invaluable tool found in many hackerspaces—and yearned for in those spaces that don't have one—that uses a laser to etch or cut out shapes in plastic and wood.

LED: Light emitting diode, a common component found in electronics projects.

leet: See *elite*.

leetspeak: Stereotypical hacker phraseology consisting mostly of numbers substituted for letters, with deliberate misspellings. For instance, *31337* stands for ELEET, or *elite*. Usually used ironically.

locksport: The art of lock-picking. Many hacker conventions hold lock-picking contests.

Make magazine: Respected DIY magazine published since 2005 by technology publisher O'Reilly.

MakerBot Thing-o-Matic: Hackerspace-favorite 3D printer, produced by Brooklyn startup MakerBot Industries. Often shortened to MakerBot.

MDF: Medium-density fiberboard, a common wood substitute used for DIY projects.

microcontroller: A tiny computer that can be programmed to accept data from sensors and switches while activating LEDs, motors, and other devices.

MIG welder: Common metalworking tool found in many hackerspace metal shops.

multimeter: Electronics test equipment capable of measuring voltage, resistance, and other variables.

open source: A philosophy that encourages the total sharing of the data and techniques of a project. For instance, an open source electronics kit might provide the source code and wiring schematics so other people could theoretically build their own.

oscilloscope: Device that allows for the visualization of electronic signals.

PCB: Printed circuit board; PCBs are the boards upon which electronic circuits are laid out.

perfboard: Perforated, nonconductive board used for laying out electronics projects.

processing: Easy-to-learn programming language favored by some hackers. It provides the underpinnings of the arduino programming language.

ReplicatorG: The interface program for 3D printers, including the ubiquitous MakerBot as well as others.

RepRap: Seminal hacker-friendly 3D printer, named for the project's goal: to be able to print another RepRap, except for electronic components.

RFID: Radio frequency identification, small tags that broadcast a weak signal when they're exposed to certain radio waves. Many hackerspaces use RFID-based security systems to safeguard their facilities.

second-wave hackerspace: European hackerspaces, inspired by first-wave hackerspaces but far more open and accepting.

servo: Motor that uses electronic feedback signals to control its mechanical position.

Skeinforge: A software application, written by Enrique Perez, for converting 3D models into G-code.

solenoid: A device for converting electricity into linear motion, kind of like a motor except pushing or pulling rather than turning.

stepper: A brushless electric motor that can divide its full rotation into a series of steps, allowing for precise positional control.

third-wave hackerspace: Most hackerspaces formed after 2007, they are inspired by second-wave hackerspaces as well as the DIY movement.

Index

SYMBOLS

A

B

C

D

E

F

G

H

I

J

K

N

O

P

Q

T

U

V

W

X

Z